Fundamentals of Electrical Engineering

FIRST EDITION

BY PRASUN BARUA

ABOUT

Welcome to "Fundamentals of Electrical Engineering"! This is a nonfiction science book designed to provide you with a solid understanding of the basic principles, concepts, and applications of electrical engineering. This book aims to guide beginners through the essential topics of this field, enabling them to lay a strong foundation for further exploration and specialization. Whether you are a student, a hobbyist, or someone interested in learning about electrical engineering, this book will serve as your reliable companion. This is the first edition of the book. Thanks for reading the book.

TABLE OF CONTENTS

Chapter	Page No.
Chapter-1: Introduction to Electrical Engineering	3
Chapter-2: Electric Circuit Analysis	21
Chapter-3: DC Circuit Analysis	24
Chapter-4: AC Circuit Analysis	36
Chapter-5: Electric and Magnetic Fields	49
Chapter-6: Basic Electronic Devices	62
Chapter-7: Power Systems	79
Chapter-8: Control Systems	103
Chapter-9: Electronic Circuits and Devices	137
Chapter-10: Introduction to Electrical Machines	149

CHAPTER-1: INTRODUCTION TO ELECTRICAL ENGINEERING

Definition and Scope

1. Definition of Electrical Engineering: Electrical engineering is a branch of engineering that deals with the study, design, development, and application of electrical and electronic systems, devices, and technologies. It encompasses a wide range of topics, including the generation, transmission, distribution, and utilization of electrical energy, as well as the design and implementation of electronic circuits and systems. Electrical engineers work on various aspects of electrical and electronic technology, from power generation and distribution systems to communication systems, control systems, and electronic devices.
2. Scope of Electrical Engineering: The scope of electrical engineering is vast and encompasses several key areas. Here are some of the main domains within the field:

a. Power Systems: Electrical engineers working in power systems are involved in the design, operation, and maintenance of electrical power generation plants, transmission and distribution networks, substations, and power utilization systems. They ensure the efficient and reliable supply of electricity to residential, commercial, and industrial sectors.

b. Electronics and Circuit Design: This area focuses on the design and analysis of electronic circuits and systems.

Electrical engineers working in this domain develop electronic devices, such as amplifiers, oscillators, digital logic circuits, and microprocessors. They also work on integrated circuits, printed circuit board (PCB) design, and the integration of electronic components into various applications.

c. Control Systems: Control systems engineering involves the design and implementation of systems that regulate and control the behavior of devices and processes. Electrical engineers working in this field develop control algorithms, design feedback systems, and utilize sensors, actuators, and microcontrollers to automate and optimize various processes, ranging from industrial automation to robotics and aerospace applications.

d. Communications and Signal Processing: Electrical engineers in this area work on the design, analysis, and optimization of communication systems. They develop technologies and techniques for transmitting and receiving signals over various mediums, such as wired and wireless networks. Signal processing deals with the manipulation, analysis, and interpretation of signals to extract information and improve system performance.

e. Renewable Energy Systems: With the increasing emphasis on sustainable energy sources, electrical engineers play a crucial role in the development and integration of renewable energy systems. They work on the design and optimization of solar power systems, wind farms, hydropower plants, and energy storage solutions, aiming to harness and utilize clean and renewable sources of energy.

f. Computer Engineering: In the realm of computer engineering, electrical engineers work on the design and development of computer hardware and systems. They are involved in the design of microprocessors, memory systems, input/output interfaces, and computer networks. They also contribute to the advancement of computer architecture, digital systems, and emerging technologies such as quantum computing.

g. Biomedical Engineering: Electrical engineers specializing in biomedical engineering apply their expertise to develop medical devices, diagnostic tools, and therapeutic systems. They work on areas such as medical imaging, bioinstrumentation, medical sensors, and prosthetic devices. Their contributions help improve healthcare outcomes and enhance the quality of life for patients.

These are just a few examples of the diverse areas within electrical engineering. The field continues to evolve rapidly, driven by advancements in technology and the increasing demand for energy-efficient and sustainable solutions. Electrical engineers play a vital role in shaping our modern world, driving innovation, and improving the way we generate, distribute, and utilize electrical energy while enabling seamless communication, automation, and technological advancements.

Historical Overview of Electrical Engineering

The development of electrical engineering has a rich and fascinating history that spans several centuries. Here is a brief historical overview of key milestones and advancements in the field:

1. Early Discoveries and the Birth of Electricity (18th Century):

- In the 18th century, scientists made significant discoveries related to electricity. Benjamin Franklin conducted experiments and formulated the concept of electrical charge and the principle of conservation of charge.
- Luigi Galvani's experiments with frog muscles and Alessandro Volta's invention of the voltaic pile led to the understanding of chemical reactions producing electricity.

2. Invention of the Electric Battery (19th Century):
   - In 1800, Alessandro Volta invented the voltaic pile, the first reliable electric battery, which produced a continuous flow of electricity. This invention laid the foundation for the practical utilization of electricity.

3. Development of Electric Telegraph and Communication (19th Century):
   - In the early 19th century, inventors such as Samuel Morse and William Cooke developed the electric telegraph, which revolutionized long-distance communication. The telegraph enabled the transmission of messages using electrical signals over long distances, significantly improving communication speed and efficiency.

4. Discovery of Electromagnetism and Electromagnetic Induction (19th Century):
   - In the early 19th century, Hans Christian Ørsted discovered the relationship between electricity and magnetism, leading to the understanding of electromagnetism.
   - Michael Faraday's experiments in the mid-19th century led to the discovery of

electromagnetic induction, demonstrating the production of electric current by moving a magnet through a coil of wire. This laid the foundation for the development of electrical generators and motors.
5. Introduction of Electric Lighting (Late 19th Century):
    - Thomas Edison's invention of the practical incandescent light bulb in 1879 marked a major milestone in electrical engineering. It provided a reliable and commercially viable solution for electric lighting, transforming the way we illuminate our surroundings.
6. Advancements in Power Systems and AC/DC Battle (Late 19th - Early 20th Century):
    - The late 19th and early 20th centuries witnessed the development of power systems for electricity generation, transmission, and distribution.
    - Nikola Tesla's contributions to alternating current (AC) systems and the development of the polyphase induction motor were crucial. This sparked the AC/DC battle between Thomas Edison, a proponent of direct current (DC), and George Westinghouse, an advocate for AC. AC eventually emerged as the dominant system for long-distance power transmission.
7. Growth of Electronics and Communication Systems (20th Century):
    - The 20th century saw rapid advancements in electronics, including the invention of the vacuum tube, which enabled the amplification and control of electrical signals.

- The invention of the transistor in 1947 by John Bardeen, Walter Brattain, and William Shockley revolutionized electronics, leading to the miniaturization of electronic devices and the birth of the semiconductor industry.
    - The development of integrated circuits (ICs) in the 1960s further advanced electronics, enabling the creation of complex electronic systems on a single chip.
8. Digital Revolution and the Internet Age (Late 20th Century):
    - The late 20th century witnessed the digital revolution, with the development of digital logic circuits, microprocessors, and computer networks.
    - The invention of the Internet and its subsequent widespread adoption transformed communication and information exchange, leading to the rapid growth of interconnected systems and technologies.
9. Renewable Energy and Sustainability (21st Century):
    - With a growing focus on sustainable energy sources, the 21st century has seen advancements in renewable energy technologies, such as solar power, wind power, and energy storage systems. Electrical engineering plays a crucial role in harnessing and integrating these clean energy sources into the power grid.

The field of electrical engineering continues to evolve at a rapid pace, with ongoing research and advancements in areas like robotics, artificial intelligence, wireless communication, power electronics, and more. These developments shape our modern world, enabling technological innovations and enhancing our quality of life.

Electrical Engineering Disciplines

Electrical engineering is a vast field that encompasses a wide range of disciplines and specialties. These disciplines focus on various aspects of electrical systems, electronics, and electromagnetism. Here are some prominent disciplines within electrical engineering:

1. Power Systems Engineering: This discipline deals with the generation, transmission, distribution, and utilization of electrical power. Power systems engineers design and analyze electrical grids, power plants, transformers, and other components to ensure reliable and efficient power supply.
2. Electronics Engineering: Electronics engineering involves the design and development of electronic circuits, devices, and systems. This discipline covers a broad spectrum, including analog electronics, digital electronics, microelectronics, integrated circuits, and semiconductor devices.
3. Control Systems Engineering: Control systems engineers work on the design and implementation of systems that regulate and control the behavior of other systems. This includes feedback control, automation, robotics, process control, and the design of controllers using techniques such as PID (Proportional-Integral-Derivative) control.
4. Telecommunications Engineering: Telecommunications engineers specialize in the transmission and reception of signals, including voice, data, and video. They design communication systems, such as wireless networks, fiber-optic

networks, satellite systems, and digital communication protocols.
5. Signal Processing: Signal processing involves the analysis, manipulation, and interpretation of signals. This discipline is essential for various applications, including audio and video processing, image and speech recognition, compression algorithms, and data analysis techniques.
6. Computer Engineering: Computer engineering merges principles of electrical engineering and computer science. Computer engineers work on the design and development of computer hardware, integrated circuits, computer systems, and the integration of hardware and software components.
7. VLSI Design: Very Large-Scale Integration (VLSI) design focuses on the design and fabrication of integrated circuits (ICs) with millions or billions of transistors. VLSI engineers use specialized tools and techniques to design complex ICs used in computer processors, memory chips, and other electronic devices.
8. Renewable Energy Systems: This discipline revolves around the generation, conversion, and integration of renewable energy sources such as solar, wind, hydro, and geothermal. Engineers in this field design and optimize renewable energy systems for maximum efficiency and sustainability.
9. Biomedical Engineering: Biomedical engineers apply engineering principles to the field of medicine and healthcare. They develop and design medical devices, imaging systems, prosthetics, and assistive technologies, and work on bioinstrumentation, biomaterials, and medical imaging.
10. Nanotechnology: Nanotechnology deals with the manipulation and control of matter on a nanoscale level. Electrical engineers specializing in

nanotechnology work on developing nanoscale devices, circuits, and systems with applications in electronics, energy, materials science, and medicine.

These disciplines represent just a fraction of the many areas within electrical engineering. Each discipline requires specialized knowledge and skills, and engineers often collaborate across disciplines to solve complex problems and advance technological innovations.

Electric Circuit Elements

Electric circuits consist of various elements that work together to control the flow of electric current. These elements can be classified into two categories: passive elements and active elements. Let's explore some of the commonly used electric circuit elements:

Passive Elements:

1. Resistors: Resistors are one of the most fundamental elements in electric circuits. They resist the flow of electric current and are characterized by their resistance, measured in ohms (Ω). Resistors are used to control the current and voltage levels in a circuit.
2. Capacitors: Capacitors store electrical energy in an electric field. They consist of two conductive plates separated by an insulating material called a dielectric. Capacitors are used to store energy, smooth voltage fluctuations, and block direct current (DC) while allowing alternating current (AC) to pass.
3. Inductors: Inductors store electrical energy in a magnetic field. They are typically made of a coil of

wire wound around a core material. Inductors resist changes in current and are used in circuits to control the flow of current, filter signals, and store energy.
4. Transformers: Transformers consist of two or more coils of wire that are magnetically coupled. They are used to step up or step down the voltage level of an alternating current (AC) signal. Transformers are commonly found in power distribution systems to convert voltage levels.

Active Elements

1. Diodes: Diodes allow the flow of current in one direction while blocking it in the opposite direction. They are used to rectify AC into DC, protect circuits from reverse voltage, and control current flow.
2. Transistors: Transistors are active electronic devices that amplify or switch electronic signals and control the flow of current. They are the building blocks of modern electronic circuits and are used in a wide range of applications, including amplifiers, oscillators, and digital logic circuits.
3. Operational Amplifiers (Op-Amps): Op-amps are integrated circuits that amplify and process electrical signals. They have high gain and are widely used in analog and digital circuits for amplification, filtering, and mathematical operations.
4. Integrated Circuits (ICs): Integrated circuits are complex circuits that contain a large number of active and passive elements fabricated on a small silicon chip. ICs are the backbone of modern electronics and are used in computers, smartphones, and a wide range of electronic devices.

These are just a few examples of electric circuit elements. Engineers use these elements and combine them in various configurations to design and build circuits for specific applications. Understanding the characteristics and behavior of these elements is crucial for analyzing and designing electric circuits.

Ohm's Law and Kirchhoff's Laws

Ohm's Law and Kirchhoff's Laws are fundamental principles in electrical engineering that describe the behavior and relationships within electric circuits. They are used to analyze and solve complex electrical circuits and are essential for understanding circuit operation. Let's explore each law in detail:

1. Ohm's Law: Ohm's Law, named after the German physicist Georg Simon Ohm, relates the current flowing through a conductor to the voltage across it and the resistance of the conductor. It can be expressed mathematically as:

$$V = I * R$$

where: V is the voltage across the conductor (in volts), I is the current flowing through the conductor (in amperes), R is the resistance of the conductor (in ohms).

Ohm's Law states that the current flowing through a conductor is directly proportional to the voltage applied across it and inversely proportional to the resistance of the conductor. In other words, if the voltage increases, the current will increase, given a constant resistance. Similarly, if the resistance increases, the current will decrease, given a constant voltage.

Ohm's Law is widely used to calculate unknown values in circuits. For example, if the voltage and resistance are known, Ohm's Law can be used to determine the current. Likewise, if the current and resistance are known, Ohm's Law can be used to calculate the voltage.

2. Kirchhoff's Laws: Kirchhoff's Laws, named after the German physicist Gustav Kirchhoff, are two fundamental principles used to analyze complex electrical circuits. These laws are based on the principles of conservation of charge and conservation of energy.

a) Kirchhoff's Current Law (KCL): KCL states that the sum of currents entering a junction (or node) in a circuit is equal to the sum of currents leaving the junction. In other words, the total current flowing into a junction is equal to the total current flowing out of the junction. Mathematically, KCL can be stated as:

$$\sum I(in) = \sum I(out)$$

KCL is based on the principle that charge is conserved in a circuit. It is used to determine unknown currents or verify the correctness of current measurements in a circuit.

Currents Entering the Node Equals Currents Leaving the Node

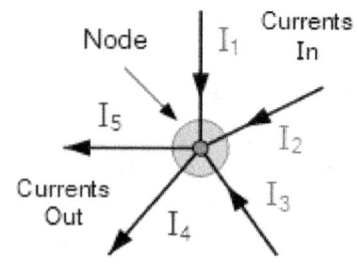

$$I_1 + I_2 + I_3 + (-I_4 + -I_5) = 0$$

b) Kirchhoff's Voltage Law (KVL): KVL states that the sum of voltages around any closed loop in a circuit is equal to zero. It is based on the principle of conservation of energy. According to KVL, the algebraic sum of the voltage rises and voltage drops around a closed loop is always zero. Mathematically, KVL can be stated as:

$$\sum V(\text{rises}) = \sum V(\text{drops})$$

KVL is used to analyze the voltages in a circuit, determine unknown voltages, or verify the correctness of voltage measurements.

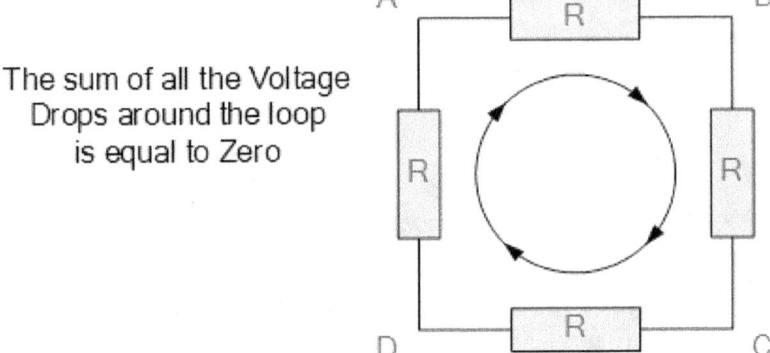

The sum of all the Voltage Drops around the loop is equal to Zero

$$V_{AB} + V_{BC} + V_{CD} + V_{DA} = 0$$

Kirchhoff's Laws are powerful tools for analyzing complex circuits that contain multiple branches and loops. By applying KCL and KVL, engineers can derive a system of equations to solve for the unknown currents and voltages in a circuit.

Overall, Ohm's Law and Kirchhoff's Laws are fundamental principles that form the basis of electrical circuit analysis. They provide a solid foundation for understanding and designing electric circuits, enabling engineers to predict and control the behavior of electrical systems.

Series and parallel circuits

Series and parallel circuits are two fundamental configurations in electrical circuits that have different characteristics and behaviors. Let's explore each circuit configuration:

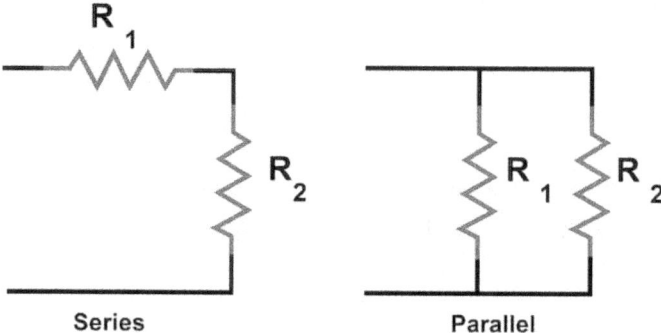

Series   Parallel

1. Series Circuits: In a series circuit, components are connected in a single pathway, forming a series loop. In this configuration, the same current flows through each component in the circuit. The key characteristics of series circuits are:

- Current: The current through each component is the same, as there is only one path for current to flow.
- Voltage: The voltage across the components in a series circuit adds up. The total voltage across the circuit is the sum of the individual voltage drops across each component.
- Resistance: The total resistance in a series circuit is the sum of the resistances of each component. The

total resistance increases as more components are added.

Key properties of series circuits

- The total resistance (R_total) is the sum of the individual resistances: R_total = R1 + R2 + R3 + ...
- The total voltage (V_total) is the sum of the individual voltage drops: V_total = V1 + V2 + V3 + ...
- The total current (I_total) is the same at all points in the circuit: I_total = I1 = I2 = I3 = ...

2. Parallel Circuits: In a parallel circuit, components are connected in multiple pathways, forming parallel branches. In this configuration, the voltage across each component is the same, while the current divides among the branches. The key characteristics of parallel circuits are:

- Current: The total current flowing into a parallel circuit is the sum of the currents flowing through each branch. The current divides among the branches based on the resistance of each branch.
- Voltage: The voltage across each component in a parallel circuit is the same. This is because all components are connected directly across the same voltage source.
- Resistance: The total resistance in a parallel circuit is inversely proportional to the sum of the reciprocals of the individual resistances. As more components are added in parallel, the total resistance decreases.

Key properties of parallel circuits:

- The total resistance (R_total) in a parallel circuit can be calculated as the reciprocal of the sum of the reciprocals of the individual resistances: 1/R_total = 1/R1 + 1/R2 + 1/R3 + ...
- The total voltage (V_total) is the same across all components: V_total = V1 = V2 = V3 = ...
- The total current (I_total) is the sum of the currents flowing through each branch: I_total = I1 + I2 + I3 + ...

Series and parallel circuits have distinct characteristics and are used in different applications. Series circuits are commonly used in situations where the same current must flow through all components, such as in Christmas tree lights or series lighting circuits. Parallel circuits, on the other hand, are used when components need to operate independently, such as in household electrical wiring or the circuitry of electronic devices.

Understanding the properties and behaviors of series and parallel circuits is essential for circuit analysis, design, and troubleshooting in electrical engineering.

Voltage and Current Division

Voltage division and current division are two fundamental concepts used in electrical circuits to determine how voltages and currents are distributed among components connected in series or parallel configurations. Let's explore each concept:

1. Voltage Division: Voltage division refers to the distribution of voltage across components connected in series. In a series circuit, the total

voltage is divided among the individual components based on their respective resistances. The voltage across each component can be calculated using the following formula:

$$V2 = (R2 / (R1 + R2)) * V\_total$$

where: V2 is the voltage across component 2, R1 and R2 are the resistances of the components, V_total is the total voltage applied across the series circuit.

According to this formula, the voltage across each component is directly proportional to its resistance. Components with higher resistance will have a greater voltage drop across them.

Voltage division is commonly used in situations where different components in a circuit require different voltage levels. For example, in a voltage regulator circuit, a resistor network can be used to divide the input voltage into different levels for various components.

2. Current Division: Current division refers to the distribution of current among components connected in parallel. In a parallel circuit, the total current is divided among the branches based on their respective resistances. The current flowing through each branch can be calculated using the following formula:

$$I2 = (R1 / (R1 + R2)) * I\_total$$

where: I2 is the current flowing through branch 2, R1 and R2 are the resistances of the branches, I_total is the total current entering the parallel circuit.

According to this formula, the current flowing through each branch is inversely proportional to its resistance. Branches with lower resistance will have a greater current flowing through them.

Current division is commonly used when multiple components in a circuit need to operate independently. For example, in a household electrical wiring system, different electrical devices are connected in parallel so that they can operate simultaneously.

Understanding voltage division and current division is crucial for circuit analysis and design. These concepts help engineers determine the behavior and performance of electrical circuits and assist in selecting appropriate component values to achieve desired voltage and current distributions.

CHAPTER-2: ELECTRICAL CIRCUIT ANALYSIS

Circuit analysis techniques

Circuit analysis techniques are essential tools used in electrical engineering to analyze and solve electrical circuits. These techniques help engineers understand the behavior of circuits, determine voltages and currents, and predict circuit performance. Let's explore some commonly used circuit analysis techniques:

1. Ohm's Law: Ohm's Law is a fundamental principle that relates voltage, current, and resistance in a circuit. It states that the current flowing through a component is directly proportional to the voltage across it and inversely proportional to its resistance. Ohm's Law (V = I * R) is used to calculate unknown voltages, currents, or resistances in a circuit.

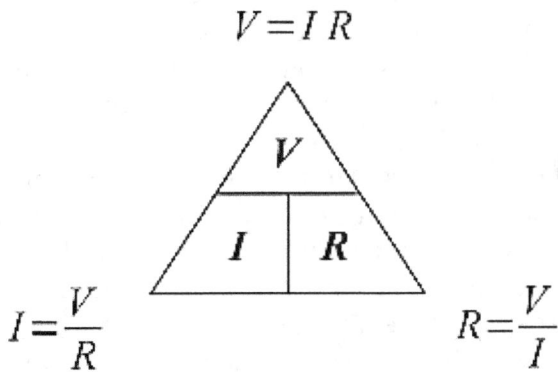

2. Kirchhoff's Laws: Kirchhoff's Laws are used to analyze complex circuits based on principles of

charge and energy conservation. These laws include:

- Kirchhoff's Current Law (KCL): KCL states that the sum of currents entering a junction in a circuit is equal to the sum of currents leaving the junction. KCL is used to determine unknown currents or verify the correctness of current measurements in a circuit.
- Kirchhoff's Voltage Law (KVL): KVL states that the sum of voltages around any closed loop in a circuit is equal to zero. KVL is used to analyze the voltages in a circuit, determine unknown voltages, or verify the correctness of voltage measurements.

By applying KCL and KVL, engineers can derive a system of equations to solve for the unknown currents and voltages in a circuit.

3. Nodal Analysis: Nodal analysis is a systematic method used to analyze circuits with multiple nodes. It involves selecting one node as a reference point and assigning voltages to the remaining nodes. By applying KCL at each non-reference node, a set of simultaneous equations is formed, which can be solved to determine the voltages at the nodes.
4. Mesh Analysis: Mesh analysis is a technique used to analyze circuits with multiple meshes or loops. It involves assigning mesh currents to each loop in the circuit and applying KVL around each mesh. By writing KVL equations for each mesh and solving the resulting system of equations, the currents in the meshes can be determined.
5. Superposition: Superposition is a method used to analyze circuits with multiple sources by considering the effect of each source individually.

The technique involves deactivating all but one source at a time and analyzing the circuit using basic circuit analysis techniques such as Ohm's Law and Kirchhoff's Laws. The responses of each individual source are then combined algebraically to obtain the final circuit response.

6. Thevenin and Norton Equivalent Circuits: The Thevenin and Norton equivalent circuits are simplification techniques used to replace complex networks with simpler representations. The Thevenin equivalent circuit represents a network as a voltage source in series with a resistor, while the Norton equivalent circuit represents a network as a current source in parallel with a resistor. These equivalents allow engineers to simplify circuit analysis and determine the behavior of a complex circuit as seen from a particular point.

These are just a few examples of circuit analysis techniques. Each technique has its advantages and is suitable for different circuit configurations. By employing these techniques, engineers can efficiently analyze and solve circuits to gain insights into circuit behavior and make informed design decisions.

CHAPTER-3: DC CIRCUIT ANALYSIS

Introduction to Direct Current (DC)

Direct Current (DC) is a form of electrical current that flows consistently in one direction. Unlike Alternating Current (AC), which periodically changes direction, DC maintains a constant polarity and magnitude over time. It is an essential concept in electrical engineering and plays a crucial role in numerous applications ranging from electronics to power transmission.

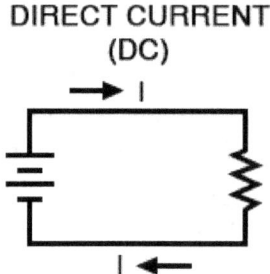

The history of DC dates back to the pioneering work of Michael Faraday and Thomas Edison in the 19th century. Faraday's experiments with electromagnetic induction laid the foundation for the generation of electrical current, while Edison's development of the practical incandescent light bulb and the distribution system relied on DC power.

In a DC circuit, electric charge flows steadily from a positive terminal to a negative terminal. This unidirectional flow is achieved through the use of power sources such as batteries, fuel cells, or rectifiers that convert AC power to DC. The flow of current is driven by the voltage potential difference between the two terminals, with electrons moving from the negative terminal (cathode) to the positive terminal (anode).

DC has several distinguishing characteristics that make it suitable for specific applications. One of its primary advantages is its ability to provide a constant and steady flow of current, making it ideal for powering devices that require a continuous supply of electricity. This characteristic is especially important in electronics, where stable voltage levels are necessary for the proper operation of integrated circuits, microprocessors, and electronic components.

Moreover, DC power is efficient for certain types of motors, such as brushed DC motors, which are commonly used in small appliances, electric vehicles, and industrial machinery. These motors rely on the consistent direction of current to generate rotational motion.

Direct current is also prevalent in renewable energy systems, particularly in photovoltaic (PV) solar panels. Solar cells generate DC electricity directly from sunlight, which can be used to power homes, buildings, or be stored in batteries for later use. In such systems, power inverters are employed to convert DC power to AC for compatibility with the electrical grid or for running AC-powered devices.

Despite its advantages, DC power transmission over long distances is challenging due to the higher energy losses encountered compared to AC transmission. Alternating current is generally preferred for long-distance power transmission because it can be easily stepped up or down using transformers, reducing losses and improving efficiency.

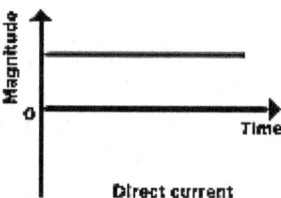

Direct current

In conclusion, Direct Current (DC) is a type of electrical current that flows in one direction and finds extensive applications in various fields. From electronics to renewable energy, DC power provides a stable and reliable source of electricity for a wide range of devices and systems. Understanding the principles and applications of DC is essential for anyone involved in electrical engineering and related disciplines.

Resistive circuits

Resistive circuits are fundamental electrical circuits consisting of resistors connected in various configurations. These circuits are based on Ohm's law, which states that the current flowing through a conductor is directly proportional to the voltage across it and inversely proportional to its resistance.

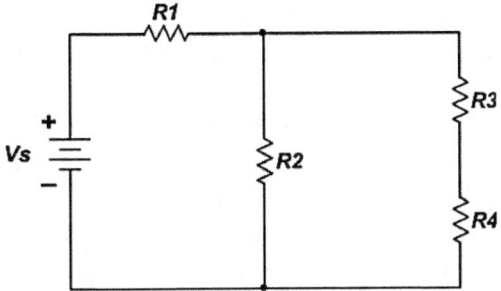

In a resistive circuit, the primary component is the resistor, which is a passive electrical device designed to impede the flow of electric current. Resistors are characterized by their resistance value, measured in ohms (Ω). They are typically constructed using materials with high resistivity, such as carbon or metal alloys.

There are two common types of resistive circuits: series circuits and parallel circuits.

1. Series Circuits: In a series circuit, resistors are connected in a sequence where the current has only one path to flow. The same current flows through each resistor, and the total resistance of the circuit is the sum of the individual resistances. Using Ohm's law, the total voltage across the series circuit can be determined by multiplying the total current by the total resistance.
2. Parallel Circuits: In a parallel circuit, resistors are connected such that they share the same voltage across their terminals. The current splits among the resistors, and the total resistance of the circuit is less than the resistance of any individual resistor. The total current flowing into the parallel circuit is the sum of the currents flowing through each resistor.

Resistive circuits exhibit certain characteristics that can be analyzed using circuit analysis techniques:

1. Current Division: In a parallel circuit, the current splits among the resistors according to their individual resistances. The current flowing through each resistor can be calculated using Ohm's law.
2. Voltage Division: In a series circuit, the voltage drops across each resistor are proportional to their resistance values. The voltage across each resistor can be determined using the voltage divider rule.
3. Power Dissipation: Resistors dissipate power in the form of heat when current flows through them. The power dissipated by a resistor can be calculated using the formula $P = I^2 * R$, where P is the power, I is the current, and R is the resistance.

Resistive circuits serve as the foundation for more complex electrical circuits and systems. They are crucial in areas such as electronics, electrical power distribution, and control systems. By understanding the behavior of resistive circuits, engineers can design and analyze more intricate circuits involving capacitors, inductors, and other components.

Overall, resistive circuits provide a fundamental understanding of how resistors interact in different configurations and play a fundamental role in electrical engineering and circuit analysis.

Nodal and Mesh Analysis

Nodal Analysis and Mesh Analysis are two widely used techniques in electrical circuit analysis that help determine voltage and current values within complex circuits. These methods are based on Kirchhoff's laws and simplify the analysis process by systematically applying these principles.

1. Nodal Analysis: Nodal Analysis, also known as Node Voltage Analysis or Kirchhoff's Current Law (KCL) Analysis, focuses on determining the unknown voltages at various nodes in a circuit. A node is a point where two or more components connect. The steps involved in nodal analysis are as follows:

   a. Identify and label the nodes in the circuit. b. Select one of the nodes as the reference or ground node and assign it a voltage of 0V. c. Apply KCL at each non-reference node by writing an equation that expresses the sum of currents entering and leaving the node as zero. d. Solve the resulting system of equations to find the voltages at all the nodes. e. Calculate the desired currents or voltages in the circuit using the node voltages.

Nodal analysis is particularly effective for circuits with many current sources, as it simplifies the process of determining unknown voltages.

2. Mesh Analysis: Mesh Analysis, also known as Loop Current Analysis or Kirchhoff's Voltage Law (KVL) Analysis, focuses on determining the unknown currents in the circuit loops or meshes. A mesh is a closed loop formed by interconnected components. The steps involved in mesh analysis are as follows:

   a. Identify and label the meshes in the circuit. b. Assign a current variable to each mesh, typically in a clockwise direction. c. Apply KVL around each mesh, writing an equation that expresses the sum of voltage drops across components in the mesh as zero. d. Solve the resulting system of equations to find the currents in each mesh. e. Calculate the desired voltages or currents in the circuit using the mesh currents.

Mesh analysis is particularly useful for circuits with many voltage sources, as it simplifies the process of determining unknown currents.

Both nodal analysis and mesh analysis provide systematic methods for solving complex electrical circuits, and the choice between them depends on the circuit's characteristics and the desired unknowns. In some cases, a combination of both techniques may be employed for more complicated circuits.

It's important to note that the application of these techniques assumes that the circuit components are linear and obey Ohm's law. Additionally, these methods are based on the assumption of ideal components and may require additional considerations for practical components such as resistors, capacitors, and inductors.

In conclusion, nodal analysis and mesh analysis are powerful tools in electrical circuit analysis, enabling engineers to determine voltage and current values in complex circuits. By applying Kirchhoff's laws and systematic analysis techniques, these methods simplify the process of solving circuit equations and provide valuable insights into circuit behavior.

Thevenin and Norton Equivalent Circuits

Thevenin and Norton Equivalent Circuits are two widely used techniques in electrical engineering that simplify complex networks by replacing them with simpler equivalent circuits. These equivalent circuits retain the same voltage-current characteristics as the original network, allowing for easier analysis and understanding. The Thevenin Equivalent Circuit is particularly useful for voltage sources, while the Norton Equivalent Circuit is more applicable for current sources.

1. Thevenin Equivalent Circuit: The Thevenin Equivalent Circuit represents a complex network as a single voltage source in series with a single equivalent resistor. This equivalent circuit can be used to simplify the analysis of the original network and determine the behavior of connected loads. The steps involved in finding the Thevenin Equivalent Circuit are as follows:

   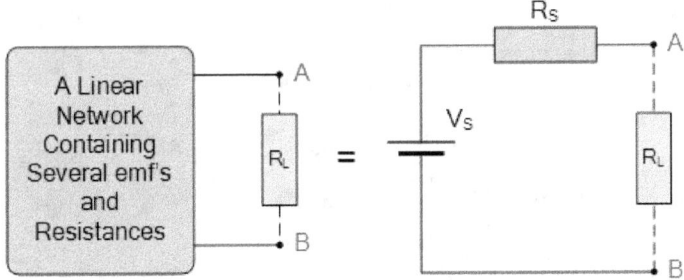

   a. Select the portion of the network that is to be replaced by its Thevenin Equivalent Circuit. b. Disconnect the load connected to the network and identify the terminals where the load was connected. c. Determine the open-circuit voltage (V_{oc}) across these terminals by calculating the voltage between them when no current is flowing. d. Find the equivalent resistance (Req) by removing all voltage and current sources from the network and calculating the resistance seen from the load terminals. e. Draw the Thevenin Equivalent Circuit, representing the open-circuit voltage as the voltage source (Vth) and the equivalent resistance as the series resistor (Rth).

The Thevenin Equivalent Circuit simplifies the analysis of complex networks, allowing for easier determination of voltage and current values for connected loads.

2. Norton Equivalent Circuit: The Norton Equivalent Circuit is similar to the Thevenin Equivalent Circuit, but it represents a complex network as a single current source in parallel with a single equivalent resistor. This equivalent circuit is particularly useful when analyzing circuits with current sources. The steps involved in finding the Norton Equivalent Circuit are as follows:

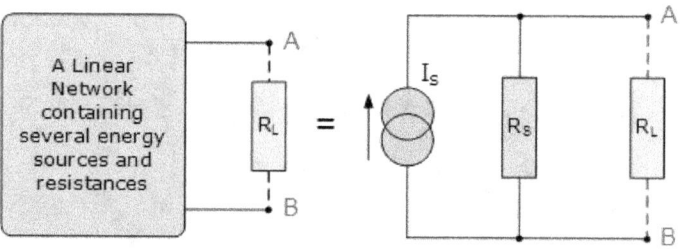

a. Select the portion of the network that is to be replaced by its Norton Equivalent Circuit. b. Disconnect the load connected to the network and identify the terminals where the load was connected. c. Determine the short-circuit current (Isc) across these terminals by calculating the current flowing through them when the terminals are shorted. d. Find the equivalent resistance (Req) by removing all voltage and current sources from the network and calculating the resistance seen from the load terminals. e. Draw the Norton Equivalent Circuit, representing the short-circuit current as the current source (In) and the equivalent resistance as the parallel resistor (Rn).

The Norton Equivalent Circuit simplifies the analysis of circuits with current sources, allowing for easier determination of voltage and current values for connected loads.

Both Thevenin and Norton Equivalent Circuits provide a simpler representation of complex networks while preserving the voltage-current characteristics of the original circuit. These equivalent circuits are invaluable in circuit analysis, allowing engineers to understand and analyze circuit behavior more efficiently. They are particularly useful in solving complex circuits with multiple interconnected elements and aid in system design and troubleshooting.

Power and Energy in DC Circuits

Power and energy are fundamental concepts in the study of electrical circuits, including DC (Direct Current) circuits. In DC circuits, the flow of electric charge is unidirectional, and power and energy play crucial roles in understanding the behavior and characteristics of these circuits.

1. Power in DC Circuits: Power is the rate at which energy is transferred or consumed in a circuit. In DC circuits, power is calculated using the formula: Power (P) = Voltage (V) × Current (I)

The unit of power is the watt (W). The watt is equivalent to one joule of energy per second. Power represents the amount of work done or energy consumed per unit of time.

2. Energy in DC Circuits: Energy is the capacity to do work, and it is associated with the stored or transferred electrical charge. In DC circuits, energy can be stored in various components like capacitors or inductors. The energy stored in a component can be calculated using different formulas based on the component's characteristics.

a. Energy in Capacitors: The energy stored in a capacitor is given by the formula: Energy (E) = 0.5 × Capacitance (C) × Voltage (V)^2

The unit of energy is the joule (J). Capacitors store energy in the electric field between their plates when charged.

b. Energy in Inductors: The energy stored in an inductor is given by the formula: Energy (E) = 0.5 × Inductance (L) × Current (I)^2

The unit of energy is also the joule (J). Inductors store energy in their magnetic fields when current flows through them.

3. Power and Energy Relationships: The relationship between power and energy is based on the concept of time. Power represents the rate at which energy is transferred or consumed. The energy consumed (or transferred) in a DC circuit can be calculated by multiplying the power by the time interval for which the power is consumed.

Energy (E) = Power (P) × Time (t)

This relationship shows that energy is the cumulative effect of power over time. For example, if a device operates at a constant power of 10 watts for 5 seconds, the energy consumed will be 50 joules.

Understanding power and energy in DC circuits is essential for analyzing and designing electrical systems. By calculating power and energy, engineers and technicians can assess the performance, efficiency, and safety of circuits, enabling them to optimize their design and operation.

CHAPTER-4: AC CURRENT ANALYSIS

Introduction to Alternating Current (AC):

Alternating Current (AC) is an important concept in the field of electrical engineering and power systems. Unlike Direct Current (DC), which flows in only one direction, AC periodically changes its direction of flow. AC is the most common form of electricity used in homes, businesses, and industries worldwide.

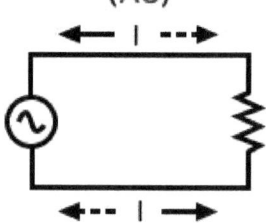

1. Basic Characteristics of AC: AC is characterized by its voltage and current that oscillate in a sinusoidal waveform. The waveform of AC is a repetitive pattern that alternates between positive and negative halves. This alternating nature of AC allows for efficient transmission and distribution of electrical power over long distances.
2. Generation of AC: AC is primarily generated by electric power stations. These power stations utilize mechanical energy, often from turbines driven by steam, water, or wind, to rotate a coil of wire within

a magnetic field. This rotation induces an alternating electromotive force (EMF) or voltage across the coil, resulting in the generation of AC.
3. Frequency and Period: The frequency of an AC waveform is the number of complete cycles it completes in one second and is measured in hertz (Hz). In most countries, the standard frequency of AC power is either 50 Hz or 60 Hz. The period is the reciprocal of the frequency and represents the time taken to complete one full cycle.
4. Voltage and Current Amplitude: In an AC waveform, the voltage and current amplitudes represent the maximum values attained during each cycle. In many applications, AC voltages are specified as root mean square (RMS) values. The RMS value of an AC waveform is equivalent to the steady DC voltage that would produce the same power in a resistive load.
5. Phases of AC: AC systems often involve multiple voltages or currents that are out of phase with each other. The phase represents the relative timing between different waveforms. In three-phase AC systems, three voltages or currents are generated, each with a phase separation of 120 degrees. Three-phase AC power is commonly used in industrial applications due to its efficiency and power distribution capabilities.
6. Advantages of AC: AC has several advantages over DC, including:

- Efficient transmission and distribution: AC can be transmitted over long distances with minimal power loss due to the ability to step-up or step-down voltages using transformers.
- Ability to change voltage levels: AC voltages can be easily transformed to different levels using

transformers, enabling efficient power distribution to various loads.
- Ease of generation: AC generation is relatively straightforward and efficient compared to generating DC.
- Reduced risk of electrical hazards: AC periodically crosses zero voltage, reducing the risk of electric shocks compared to continuous DC.

In conclusion, alternating current (AC) is the predominant form of electrical power used worldwide. Its oscillating waveform, frequency, and ability to be efficiently transmitted and distributed make it a crucial aspect of electrical systems. Understanding AC is essential for various applications, ranging from residential electricity supply to large-scale industrial power systems.

Sinusoidal waveforms

Sinusoidal waveforms are a fundamental type of waveform that are characterized by their smooth, repetitive, and oscillatory nature. These waveforms are mathematically described by the sine or cosine function and are commonly associated with alternating current (AC) signals.

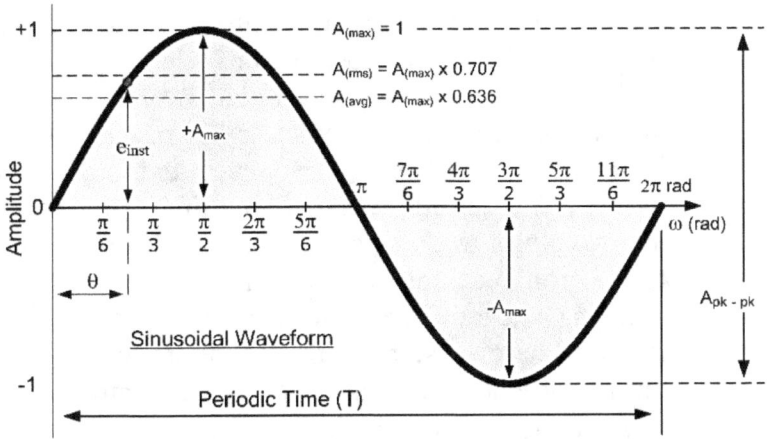

Sinusoidal Waveform

1. Shape of Sinusoidal Waveforms: A sinusoidal waveform follows the shape of a sine or cosine function, which is a smooth, continuous curve that repeats itself over regular intervals. The waveform starts from zero, rises to a positive peak, crosses zero again, reaches a negative peak, and then returns to zero. This pattern repeats indefinitely, forming a periodic waveform.
2. Amplitude: The amplitude of a sinusoidal waveform represents the maximum value reached by the waveform in either the positive or negative direction. It determines the maximum voltage or current magnitude of the waveform. The amplitude is typically denoted by "A" and can be measured from the baseline (zero) to the peak value of the waveform.
3. Frequency: The frequency of a sinusoidal waveform refers to the number of complete cycles or oscillations that the waveform completes in one second. It is measured in hertz (Hz). Frequency determines the rate at which the waveform oscillates. Higher frequencies result in more

oscillations per unit of time, while lower frequencies have fewer oscillations.
4. Period: The period of a sinusoidal waveform is the time required for one complete cycle or oscillation. It is the reciprocal of the frequency and is denoted by "T." The period represents the time it takes for the waveform to go through one full cycle, starting from zero, reaching a peak, crossing zero again, reaching a negative peak, and returning to zero.
5. Phase: Phase describes the relative position or timing of one sinusoidal waveform compared to another. It indicates the shift in time between two waveforms. Phase is often measured in degrees or radians and represents the angular displacement between corresponding points of the waveforms. Phase differences can result in constructive or destructive interference when multiple waveforms are combined.
6. Applications of Sinusoidal Waveforms: Sinusoidal waveforms find extensive applications in various fields, including:

- Alternating Current (AC) power generation, transmission, and distribution systems
- Signal processing and communication systems
- Audio and music generation
- Radio frequency (RF) and wireless communications
- Vibrations and oscillations in mechanical systems
- Harmonic analysis and Fourier series representation of complex waveforms

Sinusoidal waveforms play a crucial role in understanding and analyzing electrical circuits, signal processing, and many other areas of science and engineering. Their smooth and predictable nature allows for efficient transmission, analysis, and manipulation of signals in various applications.

Phasors and complex numbers

Phasors and complex numbers are mathematical tools used to analyze and represent sinusoidal waveforms in electrical engineering and other fields. They provide a concise and convenient way to describe the amplitude, phase, and frequency characteristics of sinusoidal signals.

1. Complex Numbers: A complex number is a number that consists of a real part and an imaginary part. It is expressed in the form a + bi, where "a" represents the real part and "b" represents the imaginary part, and "i" is the imaginary unit ($\sqrt{-1}$). Complex numbers can be added, subtracted, multiplied, and divided using specific rules.
2. Phasors: A phasor is a complex number that represents the magnitude and phase of a sinusoidal waveform. It is often used to simplify the analysis of AC circuits and signal processing. In phasor representation, the amplitude of the waveform is represented by the magnitude of the complex number, while the phase angle is represented by the argument or angle of the complex number.
3. Phasor Representation: In phasor representation, a sinusoidal waveform of the form *Acos(ωt + φ) can be represented as a complex number A*exp(jωt), where "A" is the amplitude, "ω" is the angular frequency (2πf, where "f" is the frequency), "t" is time, "j" represents the imaginary unit, and φ is the phase angle.
4. Operations with Phasors: Phasors allow for easy mathematical operations on sinusoidal waveforms. Addition and subtraction of phasors are performed

by adding or subtracting their real and imaginary parts separately. Multiplication and division of phasors are carried out by multiplying or dividing their magnitudes and adding or subtracting their phase angles.
5. Impedance: Impedance is a complex quantity that represents the opposition to the flow of alternating current in a circuit. It is the ratio of phasor voltage to phasor current. Impedance incorporates both resistance and reactance (inductive or capacitive). By using complex numbers and phasors, impedance calculations in AC circuits can be simplified.
6. Phasor Diagrams: Phasor diagrams are graphical representations of phasors. They are used to visualize the relationships between voltage and current phasors in AC circuits. Phasor diagrams help determine the magnitude, phase difference, and power relationships between different components in a circuit.

Phasors and complex numbers provide a powerful and concise mathematical framework for analyzing and solving AC circuit problems. They simplify calculations, facilitate graphical representation, and enable efficient analysis of sinusoidal waveforms, making them essential tools in electrical engineering and related disciplines.

Impedance and admittance

Impedance and admittance are concepts used in the analysis of electrical circuits, particularly in the context of alternating current (AC) circuits. They are complex quantities that describe the relationship between voltage and current in a circuit.

1. **Impedance:** Impedance, denoted by the symbol Z, is a complex quantity that represents the total opposition to the flow of AC current in a circuit. It incorporates both resistance and reactance. The impedance of a circuit is analogous to resistance in a direct current (DC) circuit.

Impedance is represented as a complex number and can be expressed as $Z = R + jX$, where R is the resistance component and X is the reactance component. The resistance (R) represents the real part of the impedance, while the reactance (X) represents the imaginary part. Reactance can be inductive (Xl) or capacitive (Xc), depending on the circuit elements involved.

The magnitude of impedance, $|Z|$, represents the total opposition to current flow, and the phase angle, φ, represents the phase relationship between voltage and current.

2. **Admittance:** Admittance, denoted by the symbol Y, is the reciprocal of impedance. It represents the ease with which AC current flows through a circuit. Admittance is also a complex quantity and is expressed as $Y = G + jB$, where G is the conductance component and B is the susceptance component.

Conductance (G) represents the real part of admittance and is analogous to the inverse of resistance. Susceptance (B) represents the imaginary part of admittance and is related to the reactance.

Similar to impedance, the magnitude of admittance, $|Y|$, represents the ease of current flow, and the phase angle, φ, represents the phase relationship between current and voltage.

3. Relationship between Impedance and Admittance: The relationship between impedance (Z) and admittance (Y) is given by the equation $Z = 1/Y$. In other words, the reciprocal of impedance is admittance, and vice versa. Both impedance and admittance describe the same circuit but from different perspectives.
4. Application: Impedance and admittance are crucial in the analysis of AC circuits. They help determine the current-voltage relationship, power distribution, resonance frequencies, and other characteristics of the circuit.

By using impedance and admittance, complex calculations involving resistance, reactance, and phase angles in AC circuits can be simplified. These concepts are extensively used in fields such as electrical engineering, electronics, telecommunications, power systems, and signal processing.

In summary, impedance and admittance are complex quantities that describe the opposition and ease of AC current flow in a circuit, respectively. They incorporate resistance and reactance and are vital for analyzing AC circuits and understanding their behavior.

AC Circuit Analysis Techniques

AC circuit analysis involves techniques and methods to analyze the behavior and characteristics of electrical circuits operating with alternating current (AC). Here are some commonly used techniques:

1. Ohm's Law for AC Circuits: Ohm's Law, which relates voltage, current, and resistance, can be extended to AC circuits by considering impedance (Z) instead of resistance (R). Ohm's Law for AC circuits states that the current (I) in an AC circuit is equal to the voltage (V) divided by the impedance (Z): $I = V/Z$.
2. Phasor Analysis: Phasor analysis is a technique that simplifies AC circuit analysis by representing sinusoidal waveforms as complex numbers called phasors. Phasors represent the amplitude and phase of the AC quantities. By converting AC voltages and currents into phasor notation, calculations involving magnitude, phase angles, and complex numbers become algebraic operations. Phasor analysis allows for straightforward calculations of voltage, current, and power in AC circuits.
3. Impedance and Admittance Analysis: Impedance (Z) and admittance (Y) are complex quantities that describe the relationship between voltage and current in AC circuits. Impedance represents the total opposition to AC current flow, while admittance represents the ease of current flow. By considering the impedance and admittance values of circuit elements (resistors, capacitors, and inductors), AC circuit analysis can be performed using complex number operations and phasor techniques.
4. Kirchhoff's Laws for AC Circuits: Kirchhoff's laws, namely Kirchhoff's Current Law (KCL) and Kirchhoff's Voltage Law (KVL), are applicable to AC circuits as well. KCL states that the sum of currents entering and leaving a node in an AC circuit is zero. KVL states that the sum of voltage rises and drops around a closed loop in an AC circuit is zero. These laws are essential tools for

analyzing complex AC circuits with multiple components and branches.
5. AC Circuit Analysis Techniques: Various techniques are used for AC circuit analysis, including:

- Mesh Analysis: Mesh analysis applies KVL to determine currents flowing in different loops (meshes) of an AC circuit. It involves writing and solving a set of simultaneous equations to find the unknown mesh currents.
- Nodal Analysis: Nodal analysis applies KCL to determine voltages at different nodes in an AC circuit. It involves writing and solving a set of simultaneous equations to find the unknown node voltages.
- Frequency Domain Analysis: Frequency domain analysis involves analyzing the behavior of AC circuits in the frequency domain. This includes using Fourier analysis techniques to decompose complex waveforms into their constituent sinusoidal components and analyzing circuit response to different frequencies.
- Network Theorems: Network theorems such as the Superposition theorem, Thevenin's theorem, and Norton's theorem are applicable to AC circuits as well. These theorems provide techniques to simplify complex circuits and determine equivalent circuits for analysis.

AC circuit analysis techniques are crucial for designing, troubleshooting, and optimizing AC electrical systems in various applications. They enable engineers to understand the voltage-current relationships, power flow, and impedance characteristics of AC circuits, facilitating efficient and reliable circuit design and operation.

Resonance and filters

Resonance and filters are important concepts in the field of electrical engineering, particularly in the analysis and design of circuits involving alternating current (AC) signals. Resonance refers to a specific frequency at which a circuit exhibits maximum response, while filters are circuits that selectively pass or reject certain frequencies.

1. Resonance: Resonance occurs in AC circuits when the frequency of the applied signal matches the natural frequency of the circuit. It leads to a significant increase in the amplitude of the response, resulting in a phenomenon called resonance. Resonance can occur in circuits containing inductors, capacitors, or a combination of both.

   - Series Resonance: In a series resonant circuit, an inductor and capacitor are connected in series. At the resonant frequency, the reactive components cancel each other out, resulting in a purely resistive impedance. This leads to a peak in the current amplitude and a minimum impedance in the circuit.
   - Parallel Resonance: In a parallel resonant circuit, an inductor and capacitor are connected in parallel. At the resonant frequency, the impedance of the circuit becomes maximum due to the combined effect of the reactive components. This results in a peak in the voltage amplitude across the circuit.

Resonance finds applications in various areas, such as tuning circuits, filters, wireless communication, and power systems.

2. Filters: Filters are circuits designed to selectively pass or attenuate specific frequency components of an input signal. They are used to shape the frequency response of a circuit and remove unwanted frequencies. Filters can be categorized into two main types: passive filters and active filters.

- Passive Filters: Passive filters are constructed using passive components like resistors, capacitors, and inductors. They do not require an external power source for their operation. Passive filters can be further classified into low-pass filters, high-pass filters, band-pass filters, and band-stop filters, depending on the frequency range they allow or reject.
- Active Filters: Active filters use active components such as operational amplifiers (op-amps) in addition to passive components. They require an external power supply to operate. Active filters offer advantages such as higher gain, better control over filter characteristics, and the ability to provide amplification in addition to filtering.

Filters play a crucial role in signal processing, audio systems, communications, image processing, and various other applications. They are used to eliminate noise, separate desired frequency components, improve signal quality, and meet specific frequency response requirements.

The design and analysis of resonant circuits and filters involve concepts from circuit theory, complex impedance, frequency response, and transfer functions. Understanding resonance and filters is essential for engineers and researchers working with AC circuits to achieve desired frequency characteristics and optimize circuit performance.

CHAPTER-5: ELECTRIC AND MAGNETIC FIELDS

Electric Fields and Coulomb's Law

fields and Coulomb's law are fundamental concepts in electrostatics, which is the study of electric charges at rest. They describe the interaction between electric charges and the creation of electric fields.

1. Electric Charges: Electric charges are fundamental properties of particles, such as protons and electrons. Charges can be positive or negative. Like charges repel each other, while opposite charges attract each other.
2. Electric Fields: An electric field is a region around a charged object or collection of charges where the influence of the electric force can be felt. Electric fields exist even in the absence of other charges, as a charged object creates an electric field around itself.
3. Coulomb's Law: Coulomb's law describes the force of attraction or repulsion between two charged objects. It states that the magnitude of the electrostatic force (F) between two point charges is directly proportional to the product of their charges (q1 and q2) and inversely proportional to the square of the distance (r) between them:

$F = k * (|q1 * q2|) / r^2$

Here, k is the electrostatic constant, also known as Coulomb's constant, which determines the strength of the electrostatic force. Its value is approximately 8.99×10^9 Nm^2/C^2.

Coulomb's law applies to point charges and assumes that the charges are at rest. It provides a mathematical relationship to calculate the force between charges based on their magnitudes and the distance between them.

4. Electric Field Strength: The electric field strength (E) at a point in space is defined as the force experienced by a unit positive test charge placed at that point. Mathematically, electric field strength is the ratio of the electric force (F) experienced by a test charge (q) to the magnitude of the test charge:

$$E = F / q$$

The electric field strength at a point is a vector quantity, meaning it has both magnitude and direction. It points in the direction of the force experienced by a positive test charge.

5. Electric Field Due to Point Charges: The electric field created by a point charge (Q) at a distance (r) from the charge can be calculated using the formula:

$$E = k * (|Q|) / r^2$$

The direction of the electric field is radial, pointing away from positive charges and towards negative charges.

Electric fields and Coulomb's law are fundamental to understanding and analyzing the behavior of electric charges and their interactions. They play a crucial role in various areas of physics, such as electromagnetism, electronics, electrical engineering, and the understanding of electric phenomena in everyday life.

Gauss's Law and Electric Flux

Gauss's law is a fundamental principle in electromagnetism that relates the electric field to the electric charge distribution. It provides a powerful tool for calculating electric fields and understanding the behavior of electric charges.

1. Gauss's Law Statement: Gauss's law states that the total electric flux through a closed surface is directly proportional to the total electric charge enclosed by that surface. Mathematically, it can be expressed as:

$$\Phi = \oint E \cdot dA = (1/\varepsilon_0) * Q$$

where Φ is the electric flux through a closed surface, E is the electric field vector, dA is an infinitesimal vector element of the surface, ε_0 is the permittivity of free space (a fundamental constant), and Q is the total charge enclosed by the surface.

2. Electric Flux: Electric flux represents the quantity of electric field passing through a given area. It is a measure of the number of electric field lines that pass through a surface. The electric flux (Φ) is defined as the dot product of the electric field vector

(E) and the area vector (dA) of an infinitesimal surface element:

$$\Phi = E \cdot dA$$

The electric flux is a scalar quantity and is typically measured in units of N·m²/C (newton-meter squared per coulomb).

3. Closed Surface and Enclosed Charge: Gauss's law relates the total electric flux through a closed surface to the total charge enclosed by that surface. A closed surface is a hypothetical surface that completely surrounds a given region of space. The charge enclosed refers to the total charge contained within that closed surface.
4. Application of Gauss's Law: Gauss's law is particularly useful in situations with high symmetry, where it allows for simplified calculations of electric fields. By choosing an appropriate Gaussian surface (a surface that matches the symmetry of the charge distribution), Gauss's law can be used to derive the electric field.

For example, if the charge distribution possesses spherical symmetry, a Gaussian surface in the form of a sphere can be chosen, and the electric field can be determined by considering the flux through that sphere.

Gauss's law is also used in the analysis of conductors and insulators, the calculation of electric field due to infinite sheets of charge, and the determination of electric fields for symmetric charge distributions, such as point charges, charged spheres, and infinite line charges.

Gauss's law, along with Coulomb's law and the principle of superposition, provides a comprehensive understanding of the behavior of electric fields and charges in electromagnetism. It is a fundamental tool for solving a wide range of problems in electrostatics and plays a significant role in the foundation of classical electromagnetism.

Electric Potential and Capacitance

Electric Potential: Electric potential, also known as voltage, is a fundamental concept in electromagnetism that describes the potential energy of a charged particle in an electric field. It represents the work done per unit charge in moving a positive test charge from a reference point to a specific location in the electric field.

1. Electric Potential Difference: The electric potential difference between two points is defined as the change in electric potential energy per unit charge between those points. Mathematically, it is given by:

$$\Delta V = V_2 - V_1 = -\int E \cdot dl$$

where ΔV is the potential difference, V_2 and V_1 are the electric potentials at the respective points, E is the electric field, and dl represents an infinitesimal displacement along the path of integration.

2. Electric Potential and Electric Field Relationship: The electric field (E) at a point is related to the electric potential (V) by the equation:

$$E = -\nabla V$$

where ∇ is the gradient operator, indicating the change in potential with respect to position.

3. Electric Potential Due to Point Charges: The electric potential (V) at a point in space due to a point charge (Q) can be calculated using the formula:

$$V = k * (Q / r)$$

where k is the electrostatic constant and r is the distance between the charge and the point at which the potential is being calculated.

4. Capacitance: Capacitance is a property of a system consisting of two conductive objects, such as parallel plates, separated by an insulating material (dielectric). It represents the ability of the system to store electric charge and is defined as the ratio of the magnitude of the charge (Q) on one object to the electric potential difference (V) between the objects:

$$C = Q / V$$

The SI unit of capacitance is the farad (F), where 1 farad is equal to 1 coulomb per volt.

5. Capacitance of Parallel Plate Capacitors: For a parallel plate capacitor, the capacitance (C) is given by:

$$C = (\varepsilon_0 * A) / d$$

where ε_0 is the permittivity of free space, A is the area of one plate, and d is the separation distance between the plates.

6. Energy Stored in a Capacitor: The energy (U) stored in a capacitor can be calculated using the formula:

$$U = (1/2) * C * V^2$$

where C is the capacitance and V is the potential difference across the capacitor.

Electric potential and capacitance are crucial concepts in understanding and analyzing electrical systems. They are utilized in various applications, including power distribution, electronic circuit design, energy storage, and many other areas of electrical engineering.

Magnetic Fields and Biot-Savart Law

Magnetic Fields: Magnetic fields are regions in space where magnetic forces are exerted on magnetic materials or moving charges. They are created by moving charges and magnets and have both magnitude and direction.

1. Magnetic Field Lines: Magnetic field lines are used to visualize and represent magnetic fields. They are drawn as continuous lines that form closed loops, indicating the direction of the magnetic field at each point. The density of the field lines represents the strength of the magnetic field, with closely spaced lines indicating a stronger field.
2. Magnetic Field Strength: The strength of a magnetic field is determined by the magnitude of the magnetic field vector (B) at a given point. The unit

of magnetic field strength is the tesla (T), where 1 tesla is equal to 1 newton per ampere-meter.
3. Biot-Savart Law: The Biot-Savart law is a fundamental principle in electromagnetism that relates the magnetic field created by a current-carrying conductor to the current and distance from the conductor. It states that the magnetic field (differential element) at a point P due to a small segment of a current-carrying wire is proportional to the current (I), the length of the wire segment (dℓ), the sine of the angle between the wire segment and the line connecting the segment to point P (θ), and inversely proportional to the square of the distance (r) between the wire segment and point P.

The Biot-Savart law equation is given as:

$$dB = (\mu_0 / 4\pi) * (I * d\ell \times \dot{A} / r^2)$$

where dB is the magnetic field vector at point P due to the wire segment, μ_0 is the permeability of free space (a fundamental constant), I is the current in the wire, dℓ is the differential length element of the wire segment, $\dot{A}$ is a unit vector representing the direction of current flow, and r is the distance between the wire segment and point P.

4. Applications of Biot-Savart Law: The Biot-Savart law is used to calculate the magnetic field produced by current-carrying wires or conductors of various shapes and configurations. It is particularly useful in determining the magnetic field around straight wires, loops, solenoids, and other simple current-carrying geometries.

By integrating the contributions of different segments of a current-carrying wire, the Biot-Savart law can be applied to find the magnetic field at any point in space due to a complex current distribution.

The Biot-Savart law is a key principle in understanding and analyzing magnetic fields and is an important tool in electromagnetic field theory and the design of devices such as transformers, motors, and generators.

It is worth noting that in cases where the currents are time-varying, additional considerations, such as Ampere's law and Maxwell's equations, need to be taken into account to fully describe the magnetic fields.

Ampere's Law and Faraday's Law

Ampere's Law and Faraday's Law are two fundamental principles in electromagnetism that describe the relationship between magnetic fields and electric currents, as well as the generation of electric fields through changing magnetic fields.

1. Ampere's Law: Ampere's Law relates the magnetic field around a closed loop to the electric currents passing through the loop. It states that the line integral of the magnetic field (B) around a closed loop is proportional to the total electric current (I) passing through the loop.

Mathematically, Ampere's Law can be stated as:

$$\oint B \cdot dl = \mu_0 * I$$

where $\oint B \cdot dl$ represents the line integral of the magnetic field around a closed loop, μ_0 is the permeability of free space, and I is the total electric current enclosed by the loop.

Ampere's Law provides a powerful tool for calculating magnetic fields due to symmetrical current distributions, such as long straight wires, circular loops, and solenoids.

2. Faraday's Law: Faraday's Law describes the relationship between changing magnetic fields and the generation of electric fields. It states that the electromotive force (EMF) induced in a closed loop is proportional to the rate of change of the magnetic flux through the loop.

Mathematically, Faraday's Law can be expressed as:

$$\varepsilon = - d\Phi / dt$$

where ε represents the induced EMF, $d\Phi/dt$ represents the rate of change of magnetic flux through the loop, and the negative sign indicates that the induced EMF opposes the change in magnetic flux.

Faraday's Law forms the basis of electromagnetic induction and explains how electric fields are generated by changing magnetic fields. It is the principle behind the functioning of devices such as generators and transformers.

3. Lenz's Law: Lenz's Law is a consequence of Faraday's Law and provides information about the direction of the induced current or the induced magnetic field. It states that the direction of the induced current or magnetic field is such that it opposes the change that produced it.

Lenz's Law ensures that the induced current or magnetic field acts to oppose the change in magnetic flux through a loop. This principle is important in understanding the conservation of energy and the behavior of electromagnetic systems.

Ampere's Law and Faraday's Law, together with Gauss's Law and the principle of conservation of charge, form the foundation of classical electromagnetism. They are essential for understanding and analyzing the behavior of electric and magnetic fields, as well as their interactions with electric currents. These laws have numerous applications in various fields, including electrical engineering, power generation, telecommunications, and electromagnetic wave propagation.

Inductance and Magnetic Circuits

Inductance: Inductance is a fundamental property of an electrical circuit that describes its ability to store energy in a magnetic field when an electric current flows through it. It is represented by the symbol L and is measured in henries (H).

1. Definition: Inductance is defined as the ratio of the magnetic flux (Φ) produced by a current-carrying coil or conductor to the current (I) flowing through it. Mathematically, it can be expressed as:

$L = \Phi / I$

The inductance of a circuit depends on various factors, including the number of turns in the coil, the shape and size of the coil, and the presence of magnetic materials.

2. Self-Inductance: Self-inductance refers to the inductance of a coil or conductor due to its own magnetic field. When the current flowing through a coil changes, it induces an electromotive force

(EMF) that opposes the change in current. This phenomenon is described by Faraday's Law of electromagnetic induction.

The self-inductance of a coil can be calculated using the formula:

$$L = (\mu_0 * N^2 * A) / \ell$$

where μ_0 is the permeability of free space, N is the number of turns in the coil, A is the cross-sectional area of the coil, and ℓ is the length of the coil.

3. Mutual Inductance: Mutual inductance refers to the inductance between two separate coils or conductors that are in close proximity to each other. When the current in one coil changes, it induces a changing magnetic field that, in turn, induces an EMF in the other coil.

The mutual inductance between two coils can be calculated using the formula:

$$M = (\mu_0 * N_1 * N_2 * A) / \ell$$

where N_1 and N_2 are the numbers of turns in the two coils, A is the common cross-sectional area, and ℓ is the separation distance between the coils.

Magnetic Circuits: Magnetic circuits are analogous to electric circuits but deal with the flow of magnetic flux instead of electric current. They are used to analyze and design magnetic systems involving magnetic materials, such as transformers, motors, and generators.

1. Magnetic Circuit Elements: Magnetic circuits consist of various elements, including magnetic cores made of ferromagnetic materials (e.g., iron), air gaps, and winding coils. These elements are analogous to resistors, capacitors, and inductors in electric circuits.
2. Magnetic Circuit Analysis: Magnetic circuit analysis involves determining the magnetic field strength (H), magnetic flux (Φ), and magnetic reluctance (R) in a magnetic circuit. The relationship between these quantities is similar to Ohm's Law in electric circuits:

$$\Phi = H * R$$

where Φ is the magnetic flux and R is the magnetic reluctance, which is a measure of the opposition to the flow of magnetic flux.

3. Magnetic Equivalent Circuits: Magnetic circuits can be represented using equivalent circuits, similar to electric circuits. These circuits use magnetic circuit elements, such as inductors and magnetic resistors, to model the behavior of magnetic systems.

By analyzing magnetic circuits, engineers can determine the magnetic field distribution, magnetic flux density, and other parameters crucial for designing efficient and optimized magnetic devices. Inductance is a property of electrical circuits that describes their ability to store energy in a magnetic field. Magnetic circuits, on the other hand, are used to analyze magnetic systems and involve the flow of magnetic flux through magnetic elements. Both concepts are important in understanding and designing magnetic systems and devices.

CHAPTER-6: BASIC ELECTRONIC DEVICES

Semiconductors and diodes

Semiconductors and diodes are fundamental components of modern electronics. Semiconductors are materials that have electrical conductivity between conductors (such as metals) and insulators (such as non-metals). Diodes, which are based on semiconductor materials, are essential devices that allow the flow of electric current in one direction while blocking it in the opposite direction. Let's explore these concepts further:

Semiconductors:

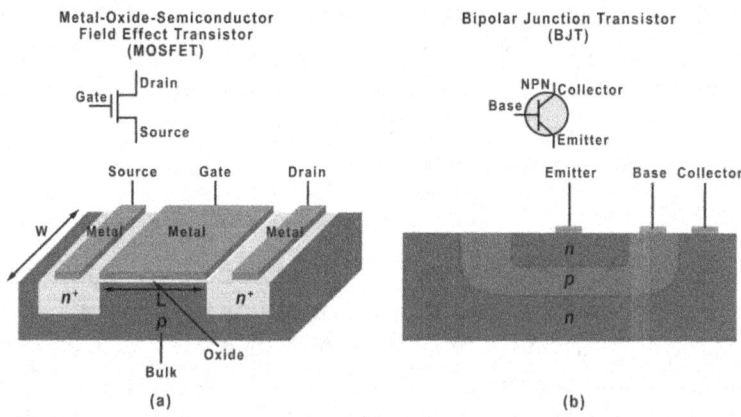

1. Semiconductor Materials: Semiconductors are typically made of elements from Group IV of the periodic table, such as silicon (Si) and germanium (Ge). These materials have four valence electrons,

making them suitable for forming a crystalline lattice structure.
2. Band Structure: The band structure of a semiconductor refers to the arrangement of energy levels or bands that electrons can occupy. Semiconductors have a valence band, which contains electrons bound to atoms, and a conduction band, which is empty or partially filled with electrons that are free to move. The energy gap between the valence and conduction bands is called the bandgap.
3. Intrinsic and Extrinsic Semiconductors: Intrinsic semiconductors are pure semiconducting materials with no intentional impurities. They have a limited number of electrons in the conduction band and holes (absence of an electron) in the valence band. Thermal excitation or the addition of energy can cause electrons to move to the conduction band, creating free carriers.

Extrinsic semiconductors are doped with impurity atoms to enhance their electrical properties. Doping introduces additional charge carriers, either electrons (n-type doping) or holes (p-type doping), into the semiconductor, thereby altering its conductivity.

Diodes:

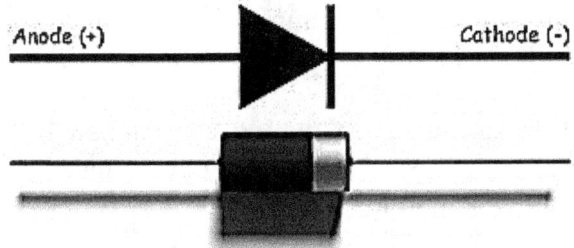

1. Diode Structure: A diode is a two-terminal electronic device formed by combining p-type and n-type semiconductors. The resulting structure is known as a p-n junction. The p-type region has an excess of positively charged holes, while the n-type region has an excess of negatively charged electrons.

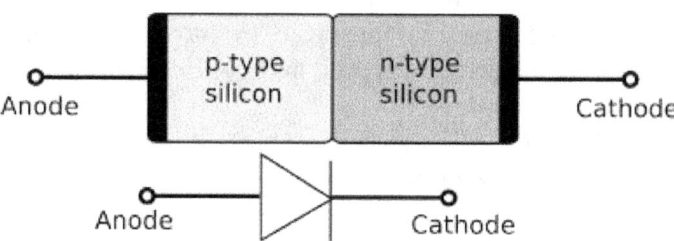

2. Diode Behavior: A diode allows electric current to flow in one direction, called forward bias, while blocking the current in the opposite direction, called reverse bias. When forward biased, the p-n junction allows electrons from the n-region to combine with holes from the p-region, creating a low resistance path for current flow. In reverse bias, the depletion region widens, preventing current flow due to the absence of available charge carriers.
3. Diode Applications: Diodes have numerous applications in electronics, including:

- Rectification: Converting AC (alternating current) to DC (direct current) by using a diode as a rectifier.
- Voltage Regulation: Stabilizing voltage levels in power supplies.
- Signal Demodulation: Extracting information from modulated signals, such as in radio communication.
- Protection: Preventing reverse current flow and protecting circuits from voltage spikes.

Diodes are available in various types, such as the standard diode, Schottky diode, Zener diode, and light-emitting diode (LED), each designed for specific applications.

Semiconductors and diodes form the foundation of modern electronics and enable the development of advanced devices and circuits. Their unique electrical properties and behaviors make them essential components in a wide range of applications, from simple rectification to complex integrated circuits and digital systems.

Bipolar Junction Transistors (BJTs)

Bipolar Junction Transistors (BJTs) are three-layer semiconductor devices that amplify or switch electronic signals. They are widely used in various electronic circuits and are classified into two types: NPN (negative-positive-negative) and PNP (positive-negative-positive) transistors. Let's explore the structure, operation, and applications of BJTs:

Structure of BJT: A BJT consists of three semiconductor layers: an emitter, a base, and a collector. In an NPN transistor, the emitter is made of N-type material, the base is made of P-type material, and the collector is made of N-type material. In a PNP transistor, the materials are reversed.

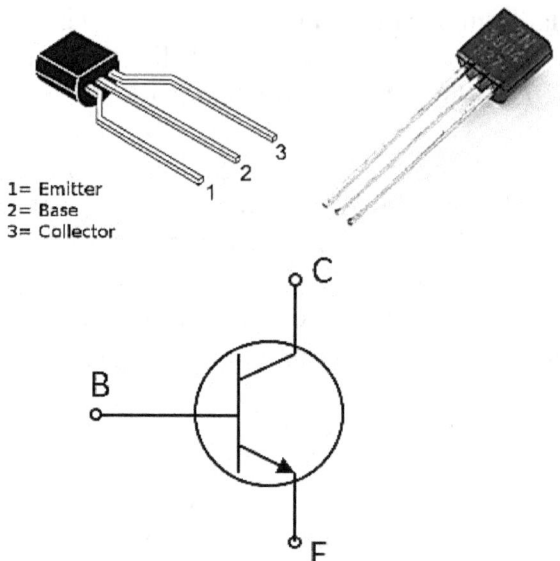

1= Emitter
2= Base
3= Collector

Operation of BJT: The operation of a BJT involves the control of current flow between the collector and emitter by the base current. BJTs have two possible modes of operation: active mode and cutoff mode.

1. Active Mode: In the active mode, the BJT is biased such that the base-emitter junction is forward-biased, allowing a current to flow from the emitter to the base. This current controls the current flow from the collector to the emitter. In the active mode, the BJT acts as an amplifier, with the output current being a multiple of the input current.
2. Cutoff Mode: In the cutoff mode, the base-emitter junction is reverse-biased, blocking the current flow from the emitter to the base. As a result, the collector current is also blocked, and the transistor is essentially off.

Applications of BJT: BJTs have numerous applications in electronics, including:

1. Amplification: BJTs are widely used as voltage and current amplifiers in analog electronic circuits. By controlling the base current, the BJT can amplify weak signals to higher power levels.
2. Switching: BJTs can be used as electronic switches to control the flow of current in electronic circuits. When in the active mode, they allow current flow, and when in the cutoff mode, they block current flow.
3. Oscillators: BJTs can be used to generate and control oscillating signals, making them useful in applications such as radio frequency (RF) oscillators and audio signal generators.
4. Digital Logic: BJTs are used in the construction of digital logic gates and memory elements in digital circuits.
5. Power Control: BJTs are used in power control circuits, such as motor drivers and power amplifiers, where high-power signals need to be controlled.

It's worth mentioning that BJTs have certain characteristics and limitations, including finite gain, temperature dependence, and voltage drop across the junctions. These factors need to be considered during the design and application of BJT-based circuits.

Overall, Bipolar Junction Transistors are versatile semiconductor devices that play a crucial role in electronic circuits for amplification, switching, and control applications. Their ability to amplify and control signals has made them essential components in various electronic devices and systems.

Field-Effect Transistors (FETs)

Field-Effect Transistors (FETs) are three-terminal semiconductor devices that use an electric field to control the flow of current. They are widely used in electronic circuits for amplification, switching, and other applications. FETs come in different types, with the two main categories being Junction Field-Effect Transistors (JFETs) and Metal-Oxide-Semiconductor Field-Effect Transistors (MOSFETs). Let's explore the structure, operation, and applications of FETs:

Structure of FET: FETs consist of three main regions: the source, the drain, and the gate. These regions are formed within a semiconductor material, typically silicon. FETs have either an N-channel or P-channel configuration, indicating the type of majority carriers (electrons or holes) that dominate the current flow.

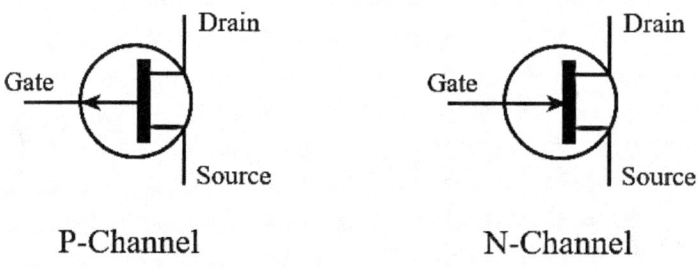

P-Channel   N-Channel

Operation of FET: The operation of FETs involves the control of current flow between the source and drain terminals by varying the voltage applied to the gate terminal. FETs work based on the modulation of the conductive channel between the source and drain regions.

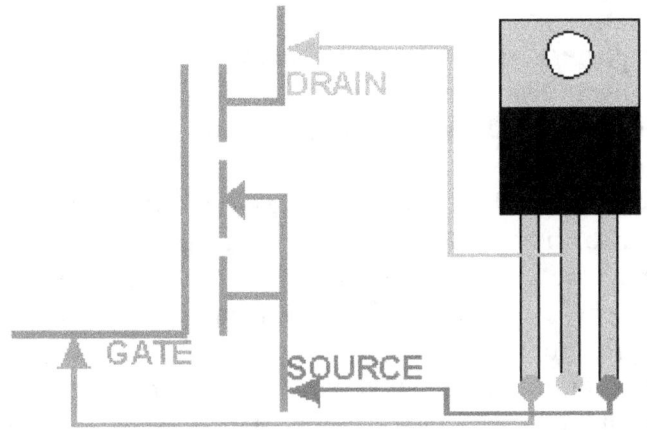

1. JFET (Junction Field-Effect Transistor): JFETs have a simple structure consisting of a channel of either N-type or P-type material, with two heavily doped regions on either side acting as the source and drain. The gate terminal is a reverse-biased PN junction that controls the width of the conducting channel. In JFETs, the current flow is controlled by the voltage applied to the gate.
2. MOSFET (Metal-Oxide-Semiconductor Field-Effect Transistor): MOSFETs have a more complex structure that includes a thin insulating layer (oxide) between the gate and channel regions. The gate terminal is separated from the channel by the insulating layer, typically made of silicon dioxide (SiO2). MOSFETs are further classified into two types: NMOS (N-channel MOSFET) and PMOS (P-channel MOSFET). The voltage applied to the gate terminal controls the formation of a conductive channel between the source and drain regions.

Applications of FETs: FETs find a wide range of applications in various electronic systems, including:

1. Amplification: FETs are commonly used as voltage amplifiers in audio and radio frequency (RF) applications. Their high input impedance and low output impedance make them suitable for amplifying weak signals.
2. Switching: FETs can be used as electronic switches due to their ability to control the flow of current. They are employed in digital logic circuits, power switches, and high-frequency switching applications.
3. Oscillators: FETs can be used in the construction of oscillators, such as RF oscillators and voltage-controlled oscillators (VCOs), for generating stable oscillating signals.
4. Analog Signal Processing: FETs are employed in various analog signal processing circuits, such as voltage regulators, filters, and audio preamplifiers.
5. Integrated Circuits: FETs are integral components of integrated circuits (ICs) and microchips, playing a key role in digital and analog circuits, memory devices, and microprocessors.

FETs offer advantages such as high input impedance, low power consumption, and faster switching speeds compared to bipolar junction transistors (BJTs). They are widely used in modern electronics due to their versatility, reliability, and scalability.

It's important to note that FETs have specific operating parameters, such as gate-source voltage limits, maximum drain-source voltage, and power dissipation, which must be considered for proper circuit design and operation.

In summary, Field-Effect Transistors (FETs) are semiconductor devices that use an electric field to control the flow of current. They offer a wide range of applications in amplification, switching, and signal processing, making them essential components in electronic circuits and systems.

Operational Amplifiers (Op-Amps)

Operational amplifiers, commonly known as op-amps, are versatile and widely used integrated circuits in electronics. They are designed to perform mathematical operations and are commonly used for amplification, signal conditioning, filtering, and other signal processing applications. Op-amps have a high gain, high input impedance, and low output impedance, making them suitable for a wide range of analog and mixed-signal circuit designs. Let's delve into the structure, characteristics, and applications of op-amps:

Structure of Op-Amps: An op-amp typically consists of a differential amplifier stage and various other circuit components. The key components of an op-amp include:

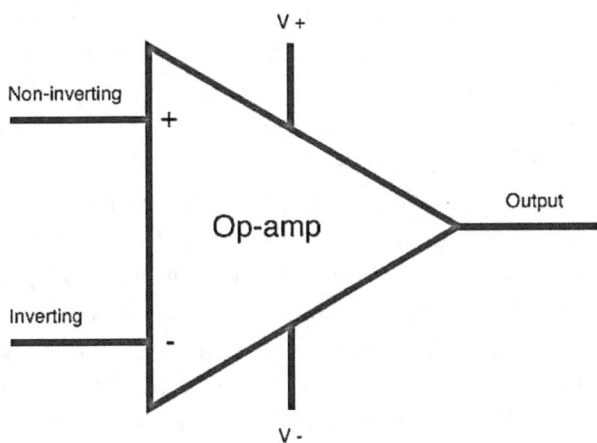

1. Differential Amplifier: This stage amplifies the voltage difference between its two input terminals, namely the non-inverting (+) and inverting (-)

inputs. The voltage gain of the differential amplifier is very high.
2. Active Loads: These loads enhance the performance and stability of the differential amplifier by providing biasing and setting the operating point.
3. Gain Stage: This stage provides additional voltage gain to the output of the differential amplifier.
4. Output Stage: The output stage of the op-amp amplifies the voltage and provides a low output impedance to drive external loads.

Characteristics of Op-Amps: Op-amps possess several important characteristics that make them valuable in circuit design:

1. High Open-Loop Gain: Op-amps have a very high open-loop voltage gain, typically in the range of tens of thousands to hundreds of thousands. This allows for precise amplification of small input signals.
2. High Input Impedance: Op-amps have a high input impedance, usually in the range of megaohms to teraohms. This ensures that the op-amp draws minimal current from the input source, thus reducing loading effects.
3. Low Output Impedance: Op-amps exhibit a low output impedance, allowing them to drive loads without significant loss of signal or voltage drop.
4. Virtual Short-circuit at Inputs: In ideal conditions, the voltage difference between the two input terminals of an op-amp is nearly zero. This virtual short-circuit property is leveraged in many op-amp circuit configurations.
5. Differential and Common-Mode Input Voltages: Op-amps have differential input voltage, which is the voltage difference between the non-inverting

and inverting inputs, as well as common-mode input voltage, which is the average voltage at both inputs.

Applications of Op-Amps: Op-amps find applications in various electronic circuits, including:

1. Amplification: Op-amps are extensively used for signal amplification in audio systems, sensors, data acquisition systems, and communication devices.
2. Filtering: Op-amps can be configured as active filters to attenuate or emphasize certain frequencies in a signal, enabling the design of low-pass, high-pass, band-pass, and notch filters.
3. Voltage and Current Regulation: Op-amps are employed in voltage regulators and current control circuits to stabilize and regulate electrical quantities.
4. Comparator: Op-amps can be used as comparators to compare two voltages and produce a digital output based on the comparison result.
5. Oscillators: Op-amps are used in the design of oscillators to generate stable and precise oscillating signals for applications such as clock circuits and waveform generation.
6. Instrumentation Amplifiers: Op-amps can be configured as instrumentation amplifiers to provide high-accuracy amplification and signal conditioning for measurement systems.

Op-amps are available in various types and configurations, each offering specific features and performance characteristics. The choice of op-amp depends on the requirements of the circuit and the desired application.

In summary, operational amplifiers (op-amps) are highly versatile integrated circuits used for amplification, signal processing, and various other applications in electronics. Their high gain, high input impedance, and low output impedance make them essential components in a wide range of analog and mixed-signal circuits.

Introduction to Digital Logic Gates

Digital logic gates are fundamental building blocks of digital circuits in the field of digital electronics. They are electronic devices that perform logical operations on one or more binary inputs and produce a binary output based on predefined logic rules. This elaboration will provide an in-depth explanation of digital logic gates, their types, and their applications in various electronic systems.

1. Basics of Digital Logic Gates:

Digital logic gates operate on binary inputs, which can be either 0 or 1, representing low voltage (logic 0) and high voltage (logic 1), respectively. The inputs and outputs of logic gates can be represented using truth tables, which define the logic relationship between inputs and outputs.

2. Types of Digital Logic Gates:

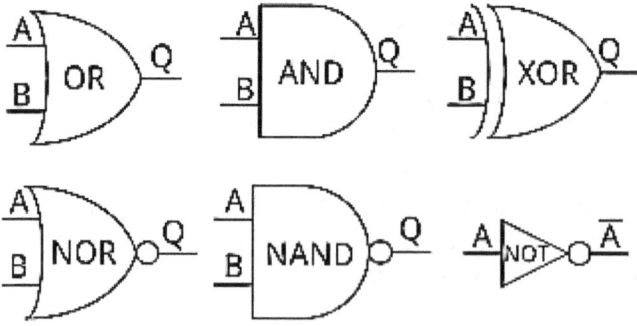

There are several types of digital logic gates, each with its specific function and truth table. The commonly used logic gates include:

- **AND Gate:** The AND gate produces a high output (logic 1) only when all of its inputs are high. The truth table for a two-input AND gate is as follows:

Input A	Input B	Output
0	0	0
0	1	0
1	0	0
1	1	1

- **OR Gate:** The OR gate produces a high output (logic 1) when any of its inputs is high. The truth table for a two-input OR gate is as follows:

Input A	Input B	Output
0	0	0
0	1	1
1	0	1
1	1	1

- **NOT Gate (Inverter):** The NOT gate, also known as an inverter, produces an output that is the complement of its input. It reverses the input logic level. The truth table for a NOT gate is as follows:

Input	Output
0	1
1	0

- **NAND Gate:** The NAND gate is an AND gate followed by a NOT gate. It produces the complement of the AND gate output. The truth table for a two-input NAND gate is as follows:

Input A Input B Output
0   0   1
0   1   1
1   0   1
1   1   0

- **NOR Gate:** The NOR gate is an OR gate followed by a NOT gate. It produces the complement of the OR gate output. The truth table for a two-input NOR gate is as follows:

Input A Input B Output
0   0   1
0   1   0
1   0   0
1   1   0

- **XOR Gate (Exclusive OR):** The XOR gate produces a high output (logic 1) when the number of high inputs is odd. The truth table for a two-input XOR gate is as follows:

Input A Input B Output
0   0   0
0   1   1
1   0   1
1   1   0

- **XNOR Gate (Exclusive NOR):** The XNOR gate produces a high output (logic 1) when the number of high inputs is even. The truth table for a two-input XNOR gate is as follows:

Input A	Input B	Output
0	0	1
0	1	0
1	0	0
1	1	1

3. Applications of Digital Logic Gates:

Digital logic gates form the foundation of digital circuits and find applications in various electronic systems:

- **Logic Circuits:** Digital logic gates are used to design and implement complex logic circuits such as adders, multiplexers, demultiplexers, flip-flops, registers, counters, and arithmetic logic units (ALUs).
- **Digital Computers:** Logic gates are the building blocks of digital computers. They are used in the central processing unit (CPU) to perform arithmetic, logic, and control operations.
- **Memory Systems:** Digital logic gates are used in memory systems to design and implement registers, memory units (RAM and ROM), and cache memory.
- **Communication Systems:** Logic gates play a vital role in designing encoding and decoding circuits, multiplexers, demultiplexers, and error detection and correction circuits used in communication systems.
- **Control Systems:** Logic gates are used to design control circuits for automation, robotics, and industrial process control applications.

- **Digital Displays:** Logic gates are used in driving digital displays, such as seven-segment displays and LCDs (Liquid Crystal Displays).
- **Security Systems:** Logic gates are used in designing security systems, such as alarm systems, access control systems, and surveillance systems.

In summary, digital logic gates are fundamental components of digital circuits in electronics. They perform logical operations on binary inputs and produce binary outputs based on predefined logic rules. Various types of logic gates exist, including AND, OR, NOT, NAND, NOR, XOR, and XNOR gates, each with its unique truth table. Digital logic gates find applications in logic circuits, digital computers, memory systems, communication systems, control systems, digital displays, and security systems. Understanding the behavior and applications of digital logic gates is essential for designing and implementing complex digital systems.

CHAPTER-7: POWER SYSTEMS

Generation, Transmission, and Distribution of electric power

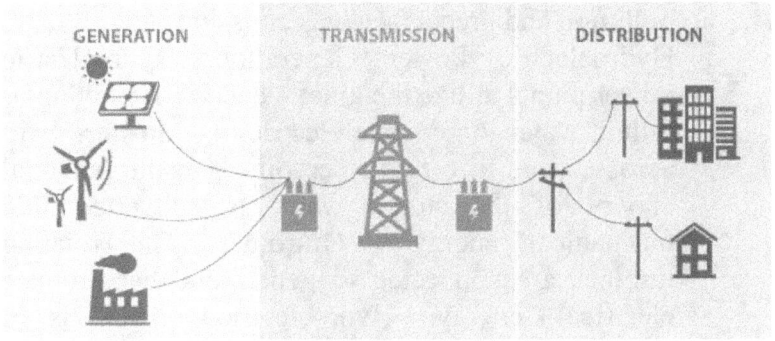

Generation, transmission, and distribution are the fundamental processes involved in delivering electric power to homes, businesses, and industries. Let's explain each of these processes in detail:

Generation

The generation of electric power is a complex process that involves the conversion of various energy sources into electrical energy. This process is vital for the functioning of modern society, as electricity powers our homes, industries, transportation systems, and more. Let's delve into the elaboration of the generation of electric power and explore some of the common methods used.

1. Thermal Power Generation: Thermal power plants are the most common and widely used method for generating electricity. These power plants use fossil fuels (coal, oil, or natural gas) or biomass as fuel to produce heat. The heat energy is then used to generate steam, which drives a turbine connected to

an electrical generator. As the turbine rotates, it converts the kinetic energy into electrical energy. Thermal power plants are known for their reliability and high energy conversion efficiency. However, they contribute to environmental issues such as air pollution and greenhouse gas emissions.
2. Hydroelectric Power Generation: Hydroelectric power plants utilize the kinetic energy of flowing or falling water to generate electricity. Dams are built across rivers to create reservoirs, and the potential energy of the stored water is converted into mechanical energy by turbines. The rotating turbines are connected to generators that produce electrical energy. Hydroelectric power is a renewable and clean energy source, as it does not produce greenhouse gas emissions during operation. However, the construction of large dams can have significant environmental and social impacts.
3. Nuclear Power Generation: Nuclear power plants harness the energy released from nuclear reactions to generate electricity. These plants utilize the process of nuclear fission, in which the nucleus of a heavy atom (such as uranium or plutonium) is split, releasing a tremendous amount of heat. This heat is used to produce steam, which drives a turbine connected to a generator. Nuclear power generation offers a high energy output and a low carbon footprint but poses challenges related to nuclear waste disposal, potential accidents, and public safety concerns.
4. Renewable Energy Generation: The increasing focus on sustainability and reducing carbon emissions has led to the rapid growth of renewable energy sources for power generation. Some of the prominent renewable energy technologies include:

- Solar Power: Photovoltaic (PV) cells convert sunlight directly into electricity. Solar power plants consist of arrays of these cells that capture solar energy and convert it into usable electrical energy.
- Wind Power: Wind turbines convert the kinetic energy of wind into mechanical energy. The rotating blades of the turbine drive a generator, producing electrical energy.
- Geothermal Power: Geothermal power plants utilize the heat from the Earth's core to generate steam, which drives turbines connected to generators.
- Biomass Power: Biomass, such as organic waste or dedicated energy crops, is burned to produce heat, which is then used to generate electricity through steam turbines.
- Tidal Power: Tidal energy converters or tidal barrages capture the energy from the rise and fall of tides to generate electricity.

5. Combined Heat and Power (CHP) Generation: Combined Heat and Power, also known as cogeneration, is a method that simultaneously produces electricity and useful heat from a single energy source. CHP systems can use various fuels such as natural gas, biomass, or waste heat from industrial processes. The generated heat can be utilized for district heating, industrial processes, or space heating, increasing overall energy efficiency.

It's worth mentioning that the electric power generation landscape is evolving, with ongoing research and development of emerging technologies such as wave power, hydrogen fuel cells, and advanced energy storage systems. These advancements aim to improve the efficiency, sustainability, and reliability of electricity generation while reducing the environmental impact.

Transmission:

Transmission of electric power is a critical process that involves the transport of electricity over long distances from power generation sources to distribution networks. It plays a crucial role in ensuring a reliable and efficient supply of electricity to homes, businesses, and industries. Let's delve into the transmission of electric power in more detail:

1. High-Voltage Transmission Lines: The backbone of the transmission system is a network of high-voltage transmission lines. These lines are designed to carry large amounts of electricity over long distances with minimal losses. High-voltage lines are typically made of aluminum or steel conductors that are capable of withstanding high voltages.
2. Step-up Transformers: Electricity generated at power plants is typically produced at lower voltages. To enable efficient transmission, step-up transformers are used to increase the voltage to higher levels, typically in the range of 115 kV to 765 kV. Increasing the voltage reduces the current flowing through the transmission lines, which helps minimize resistive losses.
3. Substations: Along the transmission network, substations act as intermediate points for voltage transformation, control, and protection. Substations receive high-voltage electricity from power plants and step it down to lower voltages suitable for distribution. These facilities house transformers,

switchgear, circuit breakers, and protective devices to ensure safe and reliable transmission.
4. Reactive Power Compensation: Reactive power is an essential component of the electrical system that ensures power quality and stability. Along transmission lines, specialized equipment such as capacitors and reactors are installed to manage reactive power. This compensation helps maintain voltage levels, reduce line losses, and improve the efficiency of the transmission system.
5. Insulation and Overhead Lines: Insulation plays a crucial role in the transmission of electric power. High-voltage transmission lines are designed with proper insulation to prevent energy losses and ensure safe operation. Insulators made of materials like porcelain or composite polymers are used to support the transmission lines and prevent electricity from escaping to the ground.
6. Tower and Pole Structures: Transmission lines are mounted on tower structures or utility poles to keep the conductors at a safe height and maintain the required separation distance between them. These structures provide mechanical support and ensure the stability of the transmission lines even in adverse weather conditions.
7. Grid Control and Monitoring: The transmission system is continuously monitored and controlled to maintain grid stability and balance the supply and demand of electricity. Supervisory control and data acquisition (SCADA) systems, along with advanced control centers, monitor various parameters such as voltage, current, and frequency in real-time. Grid operators use this information to manage the transmission system, reroute power during outages or emergencies, and prevent overloads.

8. Grid Interconnections: Transmission networks often have interconnections with neighboring regions or countries, forming a larger grid. These interconnections allow for the exchange of surplus electricity, enhance grid reliability, and enable better utilization of diverse power generation sources. Interconnection also provides backup power during emergencies and helps balance electricity supply and demand across a broader area.

Efficiency, reliability, and safety are essential considerations in the transmission of electric power. Ongoing advancements in technology and grid management techniques aim to improve the transmission infrastructure, reduce losses, enhance grid resilience, and integrate renewable energy sources effectively.

Distribution

Distribution of electric power is the final stage in the process of delivering electricity to end-users, including homes, businesses, and industries. It involves the intricate network of infrastructure that carries electricity from the transmission system and distributes it at lower voltages to individual consumers. Let's explore the distribution of electric power in more detail:

1. Distribution Substations: Distribution substations receive electricity from the high-voltage transmission lines and step down the voltage to a level suitable for local distribution. These substations are equipped with transformers that reduce the voltage from transmission voltages (e.g., 69 kV or 115 kV) to primary distribution voltages (e.g., 4 kV to 35 kV).

2. Primary Distribution Lines: Primary distribution lines carry the stepped-down voltage from the distribution substations to various areas within a region or city. These lines are typically overhead or underground cables made of aluminum or copper conductors. They are designed to handle higher currents and operate at voltages suitable for distribution to a cluster of customers.
3. Distribution Transformers: Distribution transformers are crucial components of the distribution system. They are installed at strategic locations along the primary distribution lines and further step down the voltage to levels appropriate for consumer use, typically ranging from 120/240 volts for residential customers to higher voltages for commercial and industrial customers. Transformers also provide isolation and impedance matching between the distribution network and customer premises.
4. Secondary Distribution Lines: Secondary distribution lines are responsible for delivering electricity from distribution transformers to individual consumers' premises. These lines are typically smaller in size and operate at lower voltages, such as 120/240 volts in residential areas. They are often installed overhead on utility poles or underground.
5. Service Connections: Service connections are the final link in the distribution network, connecting individual consumers to the secondary distribution lines. These connections comprise overhead or underground cables that extend from the secondary distribution lines to the metering equipment and main electrical panel in each customer's premises.
6. Metering and Switchgear: Meters are installed at customer premises to measure the amount of

electricity consumed. They provide data for billing purposes and help monitor and manage energy usage. Switchgear, including circuit breakers and other protective devices, ensures safe operation and allows for isolation and control of the electrical supply to individual premises.
7. Power Quality and Reliability: Distribution systems prioritize power quality and reliability. Power quality refers to maintaining stable voltage and frequency levels within acceptable limits. Utilities employ various measures, such as voltage regulation equipment, to ensure consistent and reliable power supply to consumers. Additionally, backup power sources, such as local generators or battery storage systems, may be deployed in critical facilities or areas prone to frequent outages.
8. Smart Grid Technologies: Modern distribution systems are incorporating advanced technologies to improve efficiency, reliability, and overall performance. Smart grid technologies enable real-time monitoring, automated fault detection, outage management, and demand response capabilities. These technologies also support the integration of renewable energy sources, energy management systems, and grid optimization algorithms.

Efficient and reliable distribution of electric power is essential to meet the demands of consumers while maintaining grid stability. Utilities continually invest in upgrading and maintaining the distribution infrastructure to ensure a robust and resilient power supply to support economic activities and enhance the quality of life.

Transformers and Power Transformers

Transformers play a crucial role in electrical power systems by facilitating the efficient transmission and distribution of electrical energy. Two common types of transformers used in power systems are transformers and power transformers. While they serve similar purposes, there are some key differences between them. Let's delve into each type in detail.

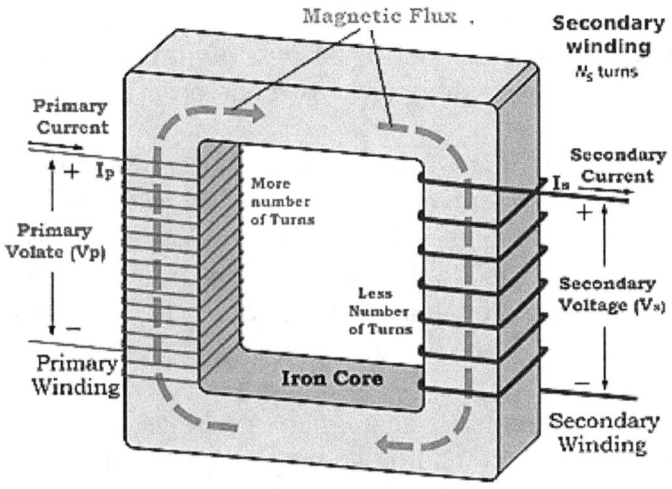

1. Transformers: Transformers, also known as distribution transformers or step-down transformers, are primarily used to step down the voltage for distribution to end-users. Here are some important features of transformers:

a. Voltage Transformation: Transformers are designed to convert high voltage levels typically used in transmission systems to lower voltage levels suitable for distribution. They operate on the principle of electromagnetic induction, where an alternating current (AC) in the primary winding creates a changing magnetic field that induces a voltage in the secondary winding.

b. Voltage Ratios: Transformers are specified by their voltage ratios. The primary winding has a higher number of turns than the secondary winding, resulting in a reduced voltage on the secondary side. The ratio of primary to secondary voltages is determined by the turns ratio.

c. Power Rating: Transformers have different power ratings, ranging from a few kVA (kilovolt-amperes) to several MVA (megavolt-amperes). The power rating depends on the amount of power the transformer can handle without exceeding its thermal and electrical limits.

d. Core and Windings: Transformers consist of a laminated iron core and two sets of windings. The primary winding is connected to the high-voltage side, while the secondary winding is connected to the low-voltage side. The windings are insulated and wound around the core to ensure efficient energy transfer.

e. Cooling Systems: Transformers generate heat during operation, so they require cooling systems to maintain optimal temperature levels. Common cooling methods include natural air cooling, forced air cooling using fans, or liquid cooling with oil or water.

f. Application: Transformers are typically used in power substations to step down the voltage for residential, commercial, and industrial use. They are responsible for supplying electricity to households, businesses, and other consumers.

2. Power Transformers: Power transformers, also known as grid transformers or step-up transformers, are primarily used in high-voltage transmission systems to step up the voltage for efficient long-

distance power transmission. Here are some important features of power transformers:

a. Voltage Transformation: Power transformers increase the voltage from a generator or a lower-voltage power system to a higher level for transmission over long distances. This higher voltage reduces transmission losses during power transfer.

b. Voltage Ratios: Similar to transformers, power transformers have specific voltage ratios. The primary winding is connected to the lower-voltage side, while the secondary winding is connected to the higher-voltage side.

c. Power Rating: Power transformers handle higher power ratings than distribution transformers. They are designed to handle several MVA of power.

d. Core and Windings: Power transformers have larger cores and windings compared to distribution transformers. They are designed to handle higher voltages and currents.

e. Cooling Systems: Due to the higher power levels involved, power transformers require efficient cooling systems. Common cooling methods include oil-immersed systems, where the windings and core are immersed in transformer oil. The oil acts as an excellent coolant and electrical insulator.

f. Application: Power transformers are located in electrical substations and facilitate the transmission of electricity over long distances. They step up the voltage at power plants to efficiently transmit electricity across the power grid.

In summary, transformers and power transformers are vital components of electrical power systems. Transformers step down the voltage for distribution, while power transformers step up the voltage for long-distance transmission. Both types are essential for ensuring efficient and reliable electricity supply to end-users.

Three-phase systems

Three-phase systems are an essential aspect of electrical engineering, widely used in power generation, transmission, and distribution. They provide a reliable and efficient means of delivering electrical energy for various applications, ranging from industrial machinery to residential power supply. This elaboration will cover the fundamental concepts, advantages, and applications of three-phase systems in electrical engineering.

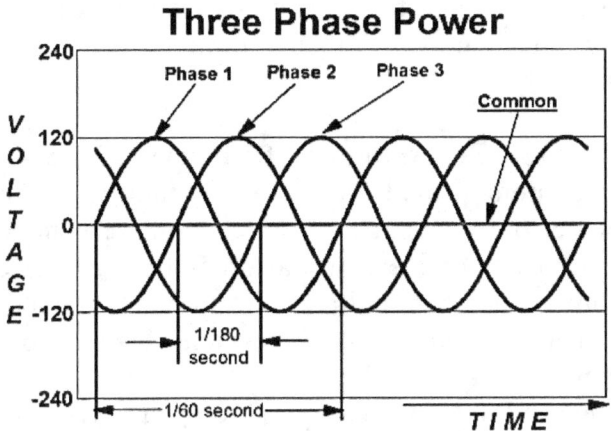

1. Basics of Three-Phase Systems:

A three-phase system consists of three sinusoidal voltage or current waveforms, which are out of phase with each other by 120 degrees. These waveforms are typically represented by the letters A, B, and C or by the phases R, Y, and B. The voltage or current in a three-phase system can be described by the following parameters:

- Line Voltage (V): The voltage between any two phases or conductors in a three-phase system.
- Phase Voltage (Vφ): The voltage between any one phase and the neutral point (if available) in a three-phase system.
- Line Current (I): The current flowing through each phase conductor in a three-phase system.
- Phase Current (Iφ): The current flowing through each phase in a three-phase system.

2. Advantages of Three-Phase Systems:

Three-phase systems offer several advantages over single-phase systems, making them preferred in various applications:

- **Power Generation Efficiency:** Three-phase generators provide a more constant and smooth power output compared to single-phase generators. The power delivered by a three-phase generator is continuous and uniform, resulting in improved efficiency and reduced mechanical stress on the generator.
- **Higher Power Capacity:** Three-phase systems can deliver significantly higher power compared to single-phase systems using the same conductor size. This is due to the balanced nature of the three-phase loads, which results in a lower current magnitude for a given power demand.
- **Balanced Loads:** In a three-phase system, loads are typically distributed evenly across the three phases. This balanced loading minimizes voltage drops and ensures equal power distribution, reducing the chances of voltage fluctuations and power loss.
- **Smaller Size and Cost:** Three-phase motors are more compact and cost-effective than their single-phase counterparts of similar power ratings. This advantage is crucial in industrial applications where large motors are frequently used.
- **Ease of Transmission:** Three-phase systems facilitate efficient power transmission over long distances. The use of higher voltages at the generation and transmission stages reduces transmission losses, allowing for cost-effective energy transfer.

3. Applications of Three-Phase Systems:

Three-phase systems find widespread applications in various fields, including:

- **Power Generation:** Most power plants and electrical grids generate and transmit electricity using three-phase systems. The generation of three-phase power by large generators provides a reliable and efficient means of delivering electricity to consumers.
- **Industrial Machinery:** Three-phase motors are extensively used in industrial applications, such as pumps, compressors, fans, conveyors, and machine tools. Their robustness, high power efficiency, and torque characteristics make them ideal for driving heavy machinery.
- **Residential Power Supply:** In residential areas, three-phase systems are commonly used to distribute power to multiple households. Although individual homes typically receive single-phase power, three-phase distribution allows for efficient allocation and balancing of the overall power demand.
- **Renewable Energy Integration:** Three-phase systems play a vital role in integrating renewable energy sources like wind and solar into the electrical grid. Large-scale wind turbines and solar power plants generate three-phase power, which can be synchronized with the grid for efficient energy transfer.
- **Electric Vehicles (EVs) Charging Stations:** With the increasing adoption of electric vehicles, three-phase power is crucial for high-power charging stations. Three-phase charging infrastructure enables faster and more efficient charging, reducing charging times for EVs.

In conclusion, three-phase systems form the backbone of modern electrical engineering. They offer numerous advantages, including higher power capacity, efficiency, and balanced loads. These systems find applications in power generation, industrial machinery, residential power supply, renewable energy integration, and electric vehicle charging stations. Understanding the principles and applications of three-phase systems is essential for electrical engineers involved in power systems design, transmission, and utilization.

Power Factor and Reactive Power

Power factor and reactive power are important concepts in electrical engineering that relate to the efficiency and quality of electrical power systems. This elaboration will provide a detailed explanation of power factor, reactive power, their significance, and their impact on power systems.

1. Power Factor:

The power factor is a dimensionless quantity that represents the ratio of real power (active power) to the apparent power in an AC electrical system. It measures the efficiency of power usage in the system. The power factor is denoted by the symbol "PF" and is calculated using the following formula:

Power Factor (PF) = Real Power (W) / Apparent Power (VA)

The real power is the portion of the power that performs useful work, such as providing energy to motors, lights, and other resistive loads. It is measured in watts (W). The apparent power is the total power supplied to the system and is a combination of real power and reactive power. It is measured in volt-amperes (VA).

The power factor can have a value between 0 and 1. A power factor of 1 indicates that the real power is equal to the apparent power, representing a system with maximum power efficiency. A power factor less than 1 indicates the presence of reactive power, which leads to a decrease in the overall power factor and energy efficiency.

2. Reactive Power:

Reactive power is a component of the apparent power that does not perform useful work but is necessary for the operation of inductive and capacitive loads in AC circuits. Inductive loads, such as motors and transformers, require reactive power to establish and maintain magnetic fields. Capacitive loads, such as capacitors, require reactive power to store and release energy.

Reactive power is measured in volt-amperes reactive (VAR). It is mathematically calculated as the product of voltage (V), current (I), and the sine of the phase angle (θ) between them:

Reactive Power (VAR) = $V * I * \sin(\theta)$

Reactive power flows back and forth between the source and the load in the AC system, causing voltage drops and additional current in the transmission and distribution lines. It does not contribute to the actual work done by the system but rather affects the power factor and system efficiency.

3. Significance of Power Factor and Reactive Power:

- **Efficiency and Energy Consumption:** A low power factor caused by reactive power results in inefficient energy usage. Reactive power increases the current required to deliver a given amount of real power, leading to increased line losses, voltage drops, and higher energy costs. Improving the power factor reduces these losses and optimizes energy consumption.
- **Load Capacity and System Voltage:** Reactive power affects the load capacity of electrical systems. High reactive power demands can overload transformers, generators, and transmission lines, limiting the overall power capacity of the system. Reactive power also affects system voltage levels, causing voltage fluctuations and instability.
- **Power Quality and Voltage Stability:** Reactive power influences the quality and stability of the voltage waveform. Insufficient reactive power can lead to voltage sags, flickering lights, and equipment malfunctions. By managing reactive power, power quality can be improved, ensuring a stable and reliable supply of electrical energy.
- **Power Factor Correction:** Power factor correction techniques aim to reduce reactive power and improve the power factor. This can be achieved through the use of power factor correction capacitors or inductors that supply or absorb reactive power, compensating for the reactive power demands of the load. Power factor correction improves energy efficiency, reduces losses, and increases the capacity of the electrical system.

4. Power Factor and Regulatory Standards:

Power factor is an important consideration in electrical systems, and utilities often enforce power factor standards.

Maintaining a high power factor reduces penalties imposed by utilities for low power factor usage. In some cases, utilities may charge customers based on their power factor to encourage power factor correction.

5. Power Factor Correction Methods:

Power factor correction can be achieved through various methods, including:

- **Capacitor Banks:** Capacitors connected in parallel with the load compensate for the reactive power demand, thereby improving the power factor.
- **Synchronous Condensers:** Synchronous motors or generators operating with no mechanical load can supply or absorb reactive power to correct the power factor.
- **Static Var Compensators (SVCs):** SVCs are solid-state devices that can provide reactive power compensation dynamically, helping to maintain a high-power factor.
- **Active Power Factor Correction (APFC):** APFC circuits use electronic devices, such as power electronics converters, to actively correct the power factor by injecting reactive power into the system.

In conclusion, power factor and reactive power are crucial aspects of electrical power systems. Power factor represents the efficiency of power usage, while reactive power is the non-working component of power required by inductive and capacitive loads. A low power factor caused by excessive reactive power leads to inefficiency, increased energy costs, and voltage instability. Power factor correction methods, such as capacitor banks and synchronous condensers, can improve power factor, reduce losses, and enhance the efficiency and stability of electrical systems.

Protection and Safety in Power Systems

Protection and safety are critical aspects of power systems to ensure the reliable and secure operation of electrical networks and to safeguard personnel and equipment. This elaboration will provide a comprehensive explanation of protection and safety measures in power systems, including their importance, key components, and various techniques employed.

1. Importance of Protection and Safety in Power Systems:

Protection and safety in power systems serve multiple purposes:

- **Equipment Protection:** Protection measures safeguard electrical equipment, such as transformers, generators, motors, and transmission lines, from abnormal operating conditions, faults, and overloads. Timely detection and isolation of faults help prevent damage to equipment, reducing downtime and maintenance costs.
- **Personnel Safety:** Safety measures protect personnel working in power systems, including operators, maintenance personnel, and the general public, from electrical hazards. They mitigate the risks associated with electric shock, arc flashes, and other electrical accidents, ensuring a safe working environment.
- **System Reliability:** Effective protection schemes enhance the reliability and stability of power systems. They isolate faulty sections promptly,

preventing cascading failures and minimizing disruptions to the supply of electrical energy.
- **Fire and Arc Flash Prevention:** Protection measures play a vital role in preventing fires and mitigating the risks associated with arc flashes. Rapid fault detection and isolation help minimize the energy released during faults, reducing the likelihood of fire and arc flash incidents.

2. Key Components of Protection and Safety Systems:

Protection and safety systems in power systems consist of several key components:

- **Relays:** Relays are the primary devices used for fault detection, isolation, and control in power systems. They sense abnormal operating conditions and send signals to circuit breakers or other protective devices to isolate faulty sections. Different types of relays are used, including overcurrent relays, differential relays, distance relays, and directional relays.
- **Circuit Breakers:** Circuit breakers are protective devices that interrupt the flow of electric current in response to relay signals. They open or close electrical circuits automatically to isolate faulty sections and prevent further damage. Circuit breakers can operate based on different principles, such as thermal, magnetic, or a combination of both.
- **Current and Voltage Transformers:** Current transformers (CTs) and voltage transformers (VTs) are used to measure currents and voltages at various points in the power system. They provide signals to relays for fault detection and protection coordination. CTs and VTs step down currents and

voltages to safer and more manageable levels for measurement and relay operation.
- **Protection Coordination:** Protection coordination involves the proper coordination and grading of protective devices in the power system. This ensures that the protective devices closest to the fault operate first, isolating the faulted section while minimizing disruption to the rest of the system. Proper coordination prevents unnecessary tripping of healthy equipment and improves the overall reliability of the protection system.
- **Grounding Systems:** Proper grounding is essential for safety in power systems. Grounding systems help dissipate fault currents and provide a reference point for voltage levels. They reduce the risk of electric shock, facilitate fault detection, and enhance system stability.
- **Backup Power Supplies:** Backup power supplies, such as uninterruptible power supply (UPS) systems and emergency generators, ensure the availability of power during outages or failures. They support critical equipment, control systems, and safety systems, maintaining operational integrity in power systems.

3. Techniques Employed for Protection and Safety:

Several techniques and methods are employed for protection and safety in power systems:

- **Overcurrent Protection:** Overcurrent protection detects excessive current flow caused by faults or overloads and isolates the faulty section. Overcurrent relays, fuses, and circuit breakers are commonly used for this purpose.

- **Differential Protection:** Differential protection compares the currents entering and leaving a protected zone or equipment. Any imbalance indicates a fault within the protected zone, triggering the operation of protective devices.
- **Distance Protection:** Distance protection measures the impedance between the relaying point and the fault location. It uses the impedance value to determine the distance to the fault and operates accordingly.
- **Transformer Protection:** Transformer protection includes measures to detect internal faults, such as winding faults, core faults, or oil insulation failures. Techniques include differential protection, Buchholz relay for oil-filled transformers, and temperature monitoring.
- **Generator Protection:** Generator protection involves techniques to detect faults within generators, such as stator faults, rotor faults, and abnormal operating conditions. Differential protection, overcurrent protection, and voltage protection are commonly employed.
- **Transmission Line Protection:** Transmission line protection includes methods to detect faults, such as short circuits and line-to-ground faults, on overhead lines and underground cables. Techniques include distance protection, overcurrent protection, and pilot relaying.
- **Arc Flash Protection:** Arc flash protection measures aim to mitigate the risks associated with arc flashes, which are hazardous electrical discharges. These measures include arc flash hazard analysis, proper personal protective equipment (PPE), and equipment design considerations.

4. Standards and Regulations:

Power system protection and safety are governed by various international and local standards and regulations. These standards provide guidelines for equipment design, protection schemes, grounding practices, arc flash mitigation, and safety procedures. Some of the widely recognized standards include those issued by the International Electrotechnical Commission (IEC), the Institute of Electrical and Electronics Engineers (IEEE), and local regulatory bodies.

In conclusion, protection and safety are vital aspects of power systems. They ensure the reliable operation of electrical networks, safeguard electrical equipment, and protect personnel from electrical hazards. Effective protection schemes employ relays, circuit breakers, transformers, coordination techniques, and grounding systems. Techniques such as overcurrent protection, differential protection, distance protection, and transformer and generator protection are used to detect and isolate faults. Compliance with standards and regulations ensures adherence to best practices and enhances the overall safety and reliability of power systems.

CHAPTER-8: CONTROL SYSTEMS

Introduction to Control Systems

Control systems are an essential part of various engineering disciplines and play a crucial role in managing and regulating the behavior of dynamic systems. Whether it's an industrial process, an aerospace application, or an electronic device, control systems are used to maintain desired operating conditions, stability, and performance. This elaboration provides a comprehensive introduction to control systems, including their definition, types, components, and applications.

1. Definition of Control Systems:

A control system is a collection of interconnected components and devices designed to regulate the behavior of a dynamic system. It compares the system's actual output with a desired or reference value, and based on the difference (error), it applies corrective measures to bring the system back to the desired state. The primary goal of a control system is to ensure stability, accuracy, efficiency, and reliability of the system's output.

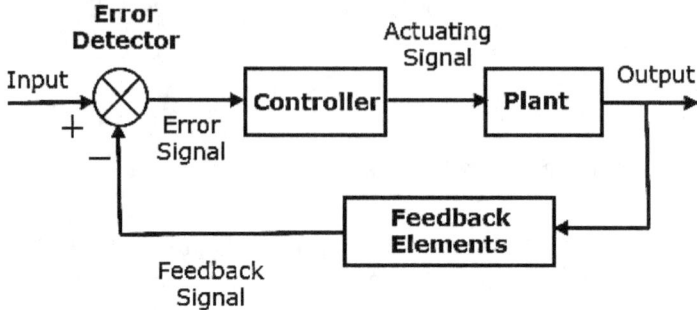

2. Types of Control Systems:

Control systems can be broadly classified into two main types:

- **Open-Loop Control System:** In an open-loop control system, the control action is not influenced by the system's output. It relies solely on the input or setpoint to generate a control signal. The system does not measure or consider the actual output or its changes. Open-loop control is typically used in situations where the system dynamics are well understood, and there is no need for feedback or correction.
- **Closed-Loop Control System:** A closed-loop (feedback) control system incorporates a feedback mechanism to continuously monitor the system's output and adjust the control action accordingly. It compares the measured output with the desired output and generates an error signal. This error signal is used to adjust the control input, ensuring that the system's output remains as close as possible to the desired value. Closed-loop control systems are widely used to improve accuracy, stability, and disturbance rejection.

3. Components of Control Systems:

Control systems consist of several key components:

- **Plant or Process:** The plant or process refers to the system or device being controlled. It can range from a simple mechanical system to a complex industrial process or an electronic circuit. The plant's behavior is influenced by external disturbances, inputs, and control actions.
- **Sensor or Measurement Device:** Sensors or measurement devices are used to measure the system's output or relevant variables. They provide feedback to the control system about the actual state or behavior of the system.
- **Controller:** The controller is the core component of a control system. It receives the measured output and the desired setpoint and computes the control action. The controller's algorithm determines how the control input is adjusted based on the error signal.
- **Actuator:** The actuator receives the control signal from the controller and converts it into physical action or energy that can affect the system. It can be a motor, a valve, a heater, or any other device that manipulates the system's inputs.
- **Feedback Path:** The feedback path establishes the connection between the system's output, the sensor, and the controller. It provides information about the system's performance and enables corrective actions based on the error signal.

4. Applications of Control Systems:

Control systems find application in numerous fields and industries:

- **Industrial Automation:** Control systems are widely used in industrial automation to regulate processes and maintain desired conditions. They control variables such as temperature, pressure, flow rate, and speed in manufacturing plants, chemical processes, power generation, and more.
- **Aerospace and Aviation:** Control systems are critical for aircraft and spacecraft. They are used for flight control, stability augmentation, navigation, and autopilot systems, ensuring safe and precise operation.
- **Robotics:** Control systems play a vital role in robotics, enabling precise motion control, trajectory tracking, object manipulation, and autonomous behavior in robotic systems.
- **Electronics and Telecommunications:** Control systems are used in electronic circuits, communication networks, and signal processing to regulate signal levels, optimize performance, and manage data flow.
- **Environmental Systems:** Control systems are employed in environmental systems such as HVAC (heating, ventilation, and air conditioning) systems, energy management systems, and pollution control systems to maintain desired environmental conditions and energy efficiency.
- **Biomedical Systems:** Control systems are utilized in medical devices, such as infusion pumps, pacemakers, and anesthesia machines, to regulate drug dosage, heart rhythms, and other critical parameters.

5. Control System Design:

Designing a control system involves several steps, including system modeling, controller design, simulation, and implementation. System modeling helps understand the system's behavior and dynamics, while controller design determines the algorithm and parameters required for optimal control. Simulation techniques allow testing and evaluating the system's performance before implementing it in the real world.

Conclusion:

Control systems are fundamental to a wide range of applications, ensuring stability, accuracy, and optimal performance of dynamic systems. Whether it's industrial automation, aerospace, robotics, or electronics, control systems play a crucial role in maintaining desired operating conditions and regulating system behavior. Understanding control system principles and components is essential for engineers working in diverse fields to design, analyze, and improve system performance.

Block Diagrams and Signal Flow Graphs

Block Diagrams and Signal Flow Graphs are graphical representations used in control system analysis and design. They provide a visual representation of the system's components and their interconnections, facilitating the understanding of the system's behavior. This elaboration will explain the concepts of Block Diagrams and Signal Flow Graphs in control systems and illustrate their construction and interpretation.

1. Block Diagrams:

Block diagrams are a graphical representation of a control system using blocks to represent system components and arrows to depict the flow of signals between them. Each block in the diagram represents a system component, such as a controller, plant, sensor, or actuator.

The following conventions are commonly used in block diagrams:

- **Summing Junction (Σ):** The summing junction is denoted by a circle and represents the algebraic sum of signals entering the junction. It is used to combine multiple signals or to represent addition or subtraction operations.
- **Transfer Function Block (G(s)):** Transfer function blocks represent the mathematical relationship between input and output signals. They are usually represented by a rectangular block with the transfer function equation inside.
- **Arrow:** Arrows indicate the direction of signal flow. The input signal is typically shown on the left side of the block, and the output signal is shown on the right side.
- **Feedback Loop:** Feedback loops represent the connection of the system's output back to a point in the control system. Feedback is crucial for closed-loop control systems.

By arranging the blocks and connecting the arrows appropriately, a block diagram provides a visual representation of the system's components and their interconnections. It allows for the analysis of signal flow, stability, and the overall behavior of the control system.

Input ---> [Controller] ---> [Plant] ---> Output

In this example, the input signal flows into the controller block, which processes the signal and sends it to the plant block. The plant block represents the system being controlled. Finally, the output signal is produced by the plant block.

2. Signal Flow Graphs:

Signal Flow Graphs (SFGs) provide another graphical representation of a control system using nodes and branches. A node represents a system variable, while a branch represents a transfer function.

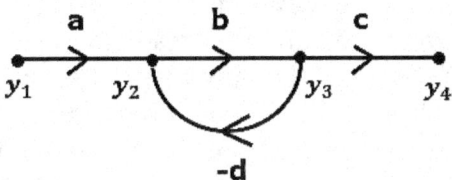

The following elements are commonly used in Signal Flow Graphs:

- **Node (Circle):** Nodes represent system variables, such as input signals, output signals, or intermediate signals within the system.
- **Branch (Directed Line):** Branches represent the flow of signals between nodes. They are directed lines with an arrow indicating the direction of signal flow.
- **Gain (Transfer Function):** The gain of a branch represents the transfer function relating the input signal to the output signal. It is typically written next to the branch.

Signal Flow Graphs offer a more compact representation compared to block diagrams and are particularly useful for analyzing large and complex control systems. They allow for the application of graph theory techniques, such as

Mason's gain formula, to compute transfer functions, system response, and stability analysis.

Interrelationship between Block Diagrams and Signal Flow Graphs:

Block diagrams and Signal Flow Graphs are interrelated representations of control systems. Given a block diagram, it is possible to derive a Signal Flow Graph, and vice versa.

To convert a block diagram to a Signal Flow Graph, each block in the block diagram corresponds to a node in the Signal Flow Graph. The arrows in the block diagram become branches in the Signal Flow Graph. The transfer functions in the block diagram are represented as gains in the Signal Flow Graph.

Conversely, to convert a Signal Flow Graph to a block diagram, each node in the Signal Flow Graph becomes a block in the block diagram. Branches with gains in the Signal Flow Graph become transfer function blocks in the block diagram.

Both representations offer insights into the system's structure and allow for analysis and design. Block diagrams are generally more intuitive and suitable for understanding the system's components, while Signal Flow Graphs provide a concise representation and lend themselves well to mathematical analysis.

Input --> Node1 --> Node2 --> Output

↑   ↑

└── Branch ──┘

In this example, the input signal flows into Node1, then to Node2, and finally results in the output signal. The branch represents the transfer function or gain between Node1 and Node2.

Conclusion:

Block diagrams and Signal Flow Graphs are graphical representations used in control system analysis and design. They provide visual representations of the system's components, signal flow, and interconnections. Block diagrams use blocks and arrows to represent system components and signal flow, while Signal Flow Graphs use nodes and branches. Understanding and utilizing both representations help engineers analyze, design, and evaluate control systems, enabling efficient control system design and optimization.

Feedback Systems and Stability in Control Systems

In control systems, feedback plays a crucial role in achieving stability, performance, and robustness. Feedback systems use information from the system's output to adjust and control the input, ensuring that the system operates as desired. This elaboration will explain the concept of feedback systems, the importance of stability, and how stability is analyzed in control systems.

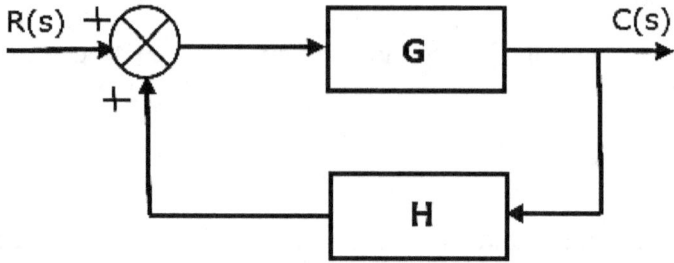

1. Feedback Systems:

A feedback system, also known as a closed-loop system, incorporates a feedback loop that connects the system's output to the input. The feedback loop allows the system to continuously monitor its output and make adjustments based on the feedback information. This enables the system to self-correct and maintain the desired behavior.

The feedback loop consists of the following components:

- **Plant or Process:** The plant refers to the system or device being controlled. It could be a physical system, such as a motor or an industrial process.
- **Sensor or Measurement Device:** The sensor measures the output of the plant and provides feedback to the control system.
- **Controller:** The controller receives the measured output and compares it to the desired setpoint. It generates a control signal or action based on the error between the desired and actual output.
- **Actuator:** The actuator receives the control signal from the controller and applies it to the plant, influencing its behavior.

The feedback loop allows the system to adjust its input based on the output, reducing errors and disturbances, improving stability, and maintaining desired performance.

2. Importance of Stability:

Stability is a fundamental characteristic of control systems. A stable system ensures that the output remains bounded and converges to a desired value in the presence of disturbances or uncertainties. Stability is crucial to prevent system oscillations, instability, or erratic behavior.

In control systems, stability is closely related to the location of the system's poles in the complex plane. Stable systems have poles with negative real parts or poles located within the unit circle for discrete-time systems.

3. Stability Analysis:

Stability analysis is the process of determining whether a control system is stable or not. It involves analyzing the system's transfer function or differential equations to assess its stability characteristics. There are several methods for stability analysis, including:

- **Routh-Hurwitz Criterion:** The Routh-Hurwitz criterion is used to determine the stability of a system by examining the coefficients of the characteristic equation.
- **Nyquist Stability Criterion:** The Nyquist criterion uses the Nyquist plot, which represents the frequency response of the system, to assess stability. It analyzes the encirclement of the critical point (-1+j0) in the complex plane.
- **Bode Stability Criterion:** The Bode criterion analyzes the frequency response of the system to

determine stability. It uses the gain and phase margins to assess stability margins.
- **Root Locus Analysis:** Root locus analysis plots the roots of the system's characteristic equation as a parameter, such as gain or pole location, is varied. It provides insights into the stability and transient response of the system.

These methods help engineers assess stability and design control systems that are stable and robust to uncertainties and disturbances.

4. Stability Criteria:

There are different stability criteria used to evaluate control systems:

- **Absolute Stability:** A system is considered absolutely stable if all its poles have negative real parts or are located within the unit circle for discrete-time systems. This ensures bounded and convergent behavior of the system.
- **Relative Stability:** Relative stability refers to the degree of stability of a system. It indicates the system's ability to handle disturbances, parameter variations, and nonlinearities without becoming unstable or exhibiting excessive oscillations.
- **Stability Margins:** Stability margins, such as gain margin and phase margin, provide measures of how close a system is to instability. They quantify the system's robustness against variations in gain and phase.

Conclusion:

Feedback systems and stability are essential concepts in control systems. Feedback enables control systems to adjust their inputs based on the output, ensuring accurate and stable operation. Stability analysis allows engineers to assess the system's stability characteristics and design control systems that are robust and reliable. Achieving stability is crucial for control system performance, disturbance rejection, and overall system behavior.

PID controllers

PID (Proportional-Integral-Derivative) controllers are widely used in control systems to regulate and stabilize dynamic processes. They are a popular choice due to their simplicity, effectiveness, and versatility. PID controllers use a combination of proportional, integral, and derivative actions to achieve desired control and response characteristics. This elaboration provides an in-depth explanation of PID controllers, including their components, working principles, tuning methods, and applications.

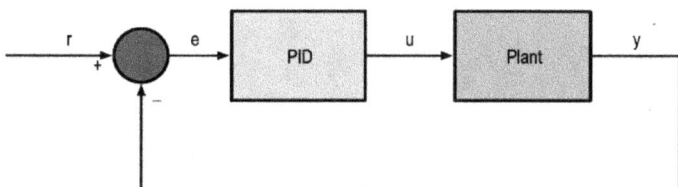

1. Components of a PID Controller:

A PID controller consists of three main components:

- **Proportional (P) Term:** The proportional term is directly proportional to the error between the desired setpoint and the measured output. It applies a control action that is proportional to the magnitude of the error. The proportional term helps reduce steady-state error and provides a response that is proportional to the current error.
- **Integral (I) Term:** The integral term accumulates the past errors over time and applies a control action

based on the integral of the error. It eliminates steady-state errors caused by systematic biases or disturbances. The integral term produces a control action that integrates the error over time and helps the system reach and maintain the desired setpoint.
- **Derivative (D) Term:** The derivative term considers the rate of change of the error with respect to time. It provides a control action that is proportional to the rate of change of the error. The derivative term helps dampen the system's response, reduce overshoot, and improve stability by anticipating future changes in the error.

The control action of a PID controller is calculated as the sum of the proportional, integral, and derivative terms, weighted by their respective gains (coefficients).

2. Working Principles of a PID Controller:

The PID controller continuously calculates the control action based on the current error and its history. The control action is applied to the system to adjust its behavior and minimize the error. The working principles of each component are as follows:

- The proportional term responds to the present error and produces an immediate control action. It provides a control effort proportional to the magnitude of the error.
- The integral term integrates the error over time, reducing steady-state errors. It continuously adjusts the control action to eliminate any long-term bias or offset.
- The derivative term considers the rate of change of the error, providing a control action that anticipates

future changes. It helps in achieving a faster response and damping oscillations.

By combining the contributions of the proportional, integral, and derivative terms, the PID controller balances the control action to achieve stability, responsiveness, and accuracy.

3. PID Controller Tuning:

Tuning a PID controller involves selecting appropriate gain values for the proportional, integral, and derivative terms to achieve desired control performance. The tuning process aims to optimize stability, response speed, and robustness. Several methods are used for PID controller tuning, including:

- **Manual Tuning:** Manual tuning involves adjusting the gain values iteratively based on system behavior and performance. It requires expert knowledge and experience to achieve satisfactory results.
- **Ziegler-Nichols Method:** The Ziegler-Nichols method is a popular heuristic tuning method. It involves step response experiments to determine critical gain and oscillation characteristics, which are used to calculate the appropriate PID gains.
- **Frequency Response Methods:** Frequency response methods use system frequency response data to design PID controller gains. These methods analyze the system's stability margins, gain and phase crossover frequencies, and shape the frequency response to achieve desired performance.

- **Auto-Tuning Algorithms:** Auto-tuning algorithms use mathematical optimization techniques or adaptive control algorithms to automatically adjust the PID gains based on system behavior and performance metrics.

The choice of tuning method depends on the system characteristics, control requirements, and available resources.

4. Applications of PID Controllers:

PID controllers find widespread applications in various industries and systems, including:

- **Industrial Processes:** PID controllers are used in industrial control systems to regulate temperature, pressure, flow rate, level, and other process variables. They are applied in manufacturing, chemical plants, power generation, and HVAC systems.
- **Robotics and Automation:** PID controllers are employed in robotic systems to control motor movements, position tracking, and stability. They help robots achieve precise and accurate control in applications such as pick-and-place operations, motion control, and robotic arms.
- **Electronics and Power Systems:** PID controllers are used in voltage regulators, power converters, inverters, and motor control systems to maintain stable and accurate output voltages and currents.
- **Biomedical Engineering:** PID controllers are utilized in medical devices such as infusion pumps, anesthesia machines, and patient monitoring

systems to ensure precise and controlled drug delivery, ventilation, and physiological parameter regulation.

Conclusion:

PID controllers are fundamental components of control systems, offering a practical and effective solution for regulating dynamic processes. By combining proportional, integral, and derivative actions, PID controllers provide stability, accuracy, and responsiveness in a wide range of applications. Understanding the components, working principles, and tuning methods of PID controllers is essential for engineers involved in control system design and optimization.

Introduction to Programmable Logic Controllers (PLCs)

Programmable Logic Controllers (PLCs) are industrial control systems that are widely used in manufacturing, automation, and process control applications. PLCs offer a flexible and reliable solution for controlling and monitoring various processes and equipment. This elaboration provides a detailed explanation of PLCs, including their components, programming languages, operation modes, and applications.

1. Components of a PLC:

A typical PLC consists of the following components:

- **Central Processing Unit (CPU):** The CPU is the brain of the PLC and is responsible for executing control logic, processing inputs and outputs, and communicating with other devices.

- **Input Modules:** Input modules interface with the external sensors and devices, converting physical signals (such as voltage or current) into digital signals that can be processed by the PLC.
- **Output Modules:** Output modules interface with actuators and devices, converting digital signals from the CPU into physical signals (such as voltage or current) that control the operation of various equipment.
- **Memory:** PLCs have different types of memory, including program memory (for storing control logic), data memory (for storing variables and values), and non-volatile memory (for storing programs and configuration even during power loss).
- **Communication Interfaces:** PLCs often include communication interfaces for connecting to other devices or systems, such as Human-Machine Interfaces (HMIs), supervisory control systems, or networks.

2. Programming Languages:

PLCs support various programming languages for developing control logic. The most common programming languages used in PLCs include:

- **Ladder Logic (LAD):** Ladder Logic is a graphical programming language that resembles electrical ladder diagrams. It uses logic symbols (contacts and coils) to represent inputs, outputs, and control functions. Ladder Logic is widely used due to its intuitive and easy-to-understand representation.
- **Function Block Diagram (FBD):** FBD is a graphical programming language that represents control logic using interconnected function blocks.

Each block performs a specific function or operation, and the interconnections define the flow of data and control signals.
- **Structured Text (ST):** Structured Text is a high-level programming language similar to structured programming languages like C or Pascal. It allows for complex programming using standard control structures like loops, conditions, and functions.
- **Sequential Function Chart (SFC):** SFC is a graphical language that combines sequential control and state transition concepts. It allows for the representation of complex sequential operations and state-based control.
- **Instruction List (IL):** Instruction List is a low-level, textual programming language that resembles assembly language. It provides direct control over the PLC's CPU and memory, making it suitable for performance-critical applications.

The choice of programming language depends on the complexity of the control logic, the familiarity of the programming team, and the specific requirements of the application.

3. Operation Modes:

PLCs can operate in different modes depending on the application requirements. The common operation modes include:

- **Run Mode:** In the Run mode, the PLC executes the control program and performs the desired control functions based on inputs, outputs, and the programmed logic.
- **Program Mode:** In the Program mode, the PLC allows users to modify, edit, or upload/download

the control program. It is used for programming, debugging, and maintenance purposes.
- **Monitor Mode:** In the Monitor mode, the PLC allows users to monitor the status of inputs, outputs, and internal variables without executing the control program. It aids in troubleshooting and verifying system behavior.
- **Remote Mode:** PLCs equipped with remote access capabilities can operate in Remote mode, enabling remote monitoring, programming, and control of the system using network connections.

4. Applications of PLCs:

PLCs are extensively used in a wide range of industries and applications, including:

- **Manufacturing Automation:** PLCs are widely used in manufacturing processes for controlling assembly lines, conveyors, robots, packaging systems, and material handling equipment.
- **Process Control:** PLCs play a vital role in process control applications such as oil refineries, chemical plants, wastewater treatment plants, and power generation facilities.
- **Building Automation:** PLCs are employed in building automation systems to control HVAC (Heating, Ventilation, and Air Conditioning) systems, lighting, access control, and security systems.
- **Food and Beverage Industry:** PLCs are utilized in food processing plants for controlling and monitoring equipment, such as mixing tanks, filling machines, and packaging lines.
- **Automotive Industry:** PLCs find applications in automotive manufacturing for controlling assembly

lines, welding robots, painting systems, and quality control processes.

Conclusion:

Programmable Logic Controllers (PLCs) are integral to modern industrial automation and control systems. They offer a flexible, reliable, and efficient solution for controlling and monitoring various processes and equipment. With their diverse programming languages, operation modes, and robust components, PLCs are extensively used in manufacturing, process control, and automation applications across various industries.

Amplifiers and Operational Amplifier Applications

Amplifiers are essential electronic devices used to increase the amplitude of an electrical signal, thereby enhancing its power, voltage, or current. They play a crucial role in a wide range of applications, from audio systems to telecommunications to instrumentation. One of the most commonly used amplifiers is the operational amplifier (op-amp), which is highly versatile and widely used in various analog circuits. This elaboration provides an in-depth explanation of amplifiers, including operational amplifiers, their characteristics, configurations, and common applications.

1. Amplifiers:

Amplifiers are electronic devices that increase the amplitude or power of an input signal while maintaining its fidelity. They are designed to provide gain, improve signal quality, drive loads, and match impedance levels. Amplifiers can be categorized based on the type of signal they amplify, such as voltage amplifiers, current amplifiers, and power amplifiers.

Amplifiers can be further classified based on their configuration, which includes common emitter, common base, common collector (in the case of transistors), and operational amplifiers (op-amps).

2. Operational Amplifiers (Op-amps):

An operational amplifier, or op-amp, is a versatile integrated circuit that is widely used in analog electronic circuits. Op-amps have high gain, high input impedance, low output impedance, and excellent linearity. They are typically available as single or dual packages and have a standardized pin configuration.

The key characteristics of op-amps include:

- **High Open-Loop Gain (Avo):** Op-amps have a very high open-loop gain, typically in the range of 10^5 to 10^6, allowing for accurate amplification of input signals.
- **Differential Inputs:** Op-amps have two input terminals, a non-inverting (+) terminal and an inverting (-) terminal. The voltage difference between these terminals determines the output.
- **Virtual Short-Circuit:** The virtual short-circuit concept states that the voltage difference between the two input terminals of an op-amp is almost zero when negative feedback is applied.
- **High Input Impedance and Low Output Impedance:** Op-amps have a high input impedance, which means they draw negligible current from the input signal source. They also have a low output

impedance, enabling them to drive low-impedance loads.
- **Supply Voltage**: Op-amps require a dual power supply or a single power supply with a virtual ground to operate correctly.

3. Op-amp Configurations:

Op-amps can be configured in various ways to suit different circuit requirements. Some common op-amp configurations include:

- **Inverting Amplifier:** In the inverting amplifier configuration, the input signal is applied to the inverting (-) terminal of the op-amp, and the amplified output is obtained from the output terminal. The input signal is inverted and amplified by a gain determined by the feedback resistor network.
- **Non-inverting Amplifier:** In the non-inverting amplifier configuration, the input signal is applied to the non-inverting (+) terminal of the op-amp, while the output is taken from the output terminal. The input signal is amplified without inversion, and the gain is determined by the feedback resistor network.
- **Differential Amplifier:** The differential amplifier configuration amplifies the voltage difference between two input signals. It is commonly used in instrumentation and communication circuits to extract the difference signal and reject common-mode noise.
- **Summing Amplifier:** A summing amplifier configuration allows multiple input signals to be summed together. It finds applications in audio

mixers, analog computing, and signal processing circuits.
- **Integrator and Differentiator:** Op-amps can be used in integrator and differentiator configurations to perform mathematical operations on the input signal. The integrator produces an output signal proportional to the integral of the input signal, while the differentiator produces an output signal proportional to the derivative of the input signal.

4. Operational Amplifier Applications:

Operational amplifiers find extensive applications in various electronic circuits, including:

- **Signal Conditioning:** Op-amps are used for signal conditioning tasks such as amplification, filtering, and impedance matching in sensor interfaces, data acquisition systems, and instrumentation.
- **Active Filters:** Op-amps are used to design active filters that provide selective frequency response for applications like audio equalizers, low-pass filters, high-pass filters, band-pass filters, and notch filters.
- **Voltage and Current Amplification:** Op-amps are used to amplify voltage and current signals in audio amplifiers, voltage regulators, motor drivers, and power supply circuits.
- **Comparators:** Op-amps can be used as comparators to compare input voltages and provide digital outputs based on specific threshold levels. They are used in analog-to-digital converters, voltage detectors, and window comparators.
- **Oscillators:** Op-amps can be configured as oscillators to generate periodic waveforms, such as square waves, sine waves, and triangular waves.

They find applications in clock generators, frequency generators, and waveform generators.
- **Voltage References:** Op-amps are used to provide stable and precise voltage references for analog circuits and systems, such as voltage regulators, analog-to-digital converters, and precision measurement equipment.

Conclusion:

Amplifiers, including operational amplifiers (op-amps), are fundamental components of analog electronic circuits. They enable signal amplification, conditioning, and processing for a wide range of applications across various industries. Op-amps, with their high gain, versatile configurations, and excellent characteristics, provide flexibility and precision in designing analog circuits for amplification, filtering, computation, and signal generation. Understanding the principles, configurations, and applications of amplifiers, including op-amps, is essential for engineers involved in analog circuit design and analysis.

Oscillators and signal generators

Oscillators and signal generators are electronic circuits used to generate continuous or periodic waveforms of specific frequencies and characteristics. They are essential components in various applications, including communication systems, signal processing, test and measurement equipment, and audio devices. This elaboration provides a detailed explanation of oscillators and signal generators, including their working principles, types, and common applications.

1. Oscillators:

Oscillators are electronic circuits that generate continuous or periodic waveforms without the need for an external input signal. They provide a stable and precise output frequency, which is essential in many applications. The key components of an oscillator circuit include an amplifier, a frequency-selective network (resistor-capacitor, inductor-capacitor, or crystal), and a feedback loop.

Working Principle of Oscillators:

Oscillators work based on the principle of positive feedback. The amplifier in the circuit amplifies the signal, and the frequency-selective network determines the desired frequency. The output of the circuit is fed back to the input with a phase shift of 180 degrees (inverting feedback) or 0 degrees (non-inverting feedback), which sustains the oscillation.

Types of Oscillators:

- **LC Oscillators:** LC oscillators use inductors and capacitors in the feedback network to generate oscillations. They are commonly used in radio frequency (RF) applications.
- **RC Oscillators:** RC oscillators use resistors and capacitors in the feedback network. They are simple and inexpensive but typically offer lower frequency stability compared to LC oscillators.
- **Crystal Oscillators:** Crystal oscillators use quartz crystals as frequency-selective elements. They offer high frequency stability and precision and are widely used in timekeeping devices, communication systems, and digital circuits.
- **Voltage-Controlled Oscillators (VCOs):** VCOs are oscillators whose frequency can be controlled by an external voltage. They are used in frequency

synthesis, phase-locked loops (PLLs), and modulation/demodulation applications.
- **Phase-Locked Loop (PLL) Oscillators:** PLL oscillators use a feedback mechanism to synchronize the output frequency with a reference frequency. They are widely used in frequency synthesis, clock generation, and communication systems.
- **Relaxation Oscillators:** Relaxation oscillators generate waveforms with a specific frequency and amplitude by charging and discharging capacitors or other energy storage elements. They are commonly used in timing circuits, waveform generation, and audio applications.

2. Signal Generators:

Signal generators are electronic devices used to produce specific waveforms or signals for testing, measurement, calibration, and simulation purposes. They provide a controlled and accurate representation of various types of signals.

Types of Signal Generators:

- **Function Generators:** Function generators are versatile signal generators that can generate different waveforms such as sine, square, triangular, and sawtooth waves. They offer adjustable frequency, amplitude, and duty cycle.
- **Arbitrary Waveform Generators (AWGs):** AWGs are advanced signal generators that can generate complex waveforms with user-defined shapes and characteristics. They are used in signal processing, communication system testing, and simulation applications.

- **Pulse Generators:** Pulse generators generate pulses with specific pulse widths, amplitudes, and repetition rates. They are used in digital circuits testing, timing analysis, and communication systems.
- **RF Signal Generators:** RF signal generators are specialized signal generators used in RF and wireless communication applications. They generate continuous wave (CW) signals, modulated signals (such as amplitude modulation (AM), frequency modulation (FM), and phase modulation (PM)), and various modulation formats used in digital communication.
- **Audio Signal Generators:** Audio signal generators are used in audio equipment testing and calibration. They generate audio signals at specific frequencies and amplitudes to test audio systems, speakers, and audio amplifiers.

Applications of Oscillators and Signal Generators:

- **Communication Systems:** Oscillators and signal generators play a vital role in communication systems for generating carrier signals, modulating and demodulating signals, and frequency synthesis.
- **Test and Measurement:** Oscillators and signal generators are extensively used in test and measurement applications for calibrating instruments, testing circuit responses, and generating specific waveforms for analysis and validation.
- **Audio Systems:** Signal generators are used in audio system testing and evaluation, speaker and amplifier calibration, and sound synthesis.
- **Research and Development:** Oscillators and signal generators are essential tools in research and

development laboratories for conducting experiments, prototyping circuits, and analyzing system behavior.
- **Electronic Instrumentation:** Oscillators and signal generators are used in electronic instrumentations such as oscilloscopes, spectrum analyzers, and frequency counters for signal generation and analysis.

Conclusion:

Oscillators and signal generators are fundamental components in electronics, providing accurate and controlled waveforms for a wide range of applications. Oscillators generate continuous or periodic waveforms, while signal generators produce specific signals for testing, measurement, and simulation purposes. Understanding the working principles, types, and applications of oscillators and signal generators is essential for engineers and technicians working in fields such as communications, test and measurement, audio systems, and research and development.

Filters and Frequency Response

Filters are electronic circuits that modify the amplitude and/or phase characteristics of a signal with respect to its frequency. They are widely used in various applications, including audio systems, communication systems, image processing, and instrumentation. Filters allow the passage of desired frequencies while attenuating or blocking unwanted frequencies. This elaboration provides a detailed explanation of filters, their types, and the concept of frequency response.

1. Types of Filters:

Filters can be classified into several types based on their frequency response characteristics and design techniques. The common types of filters include:

- **Passive Filters:** Passive filters are composed of passive components such as resistors, capacitors, and inductors. They do not require an external power source and are relatively simple in design. Passive filters can be further categorized into low-pass, high-pass, band-pass, and band-stop filters.
- **Active Filters:** Active filters incorporate active components, typically operational amplifiers (op-amps), in addition to passive components. Active filters provide higher flexibility, improved performance, and adjustable gain compared to passive filters.
- **Analog Filters:** Analog filters process continuous-time signals using analog circuitry. They are commonly used in audio applications, analog communication systems, and instrumentation.
- **Digital Filters:** Digital filters operate on discrete-time signals and are implemented using digital signal processing (DSP) techniques. They find extensive applications in digital communication systems, audio processing, and image processing.
- **FIR Filters:** Finite Impulse Response (FIR) filters have a finite impulse response, meaning their output is determined by a finite number of previous input samples. FIR filters offer linear phase response and are widely used in applications where phase distortion is critical.
- **IIR Filters:** Infinite Impulse Response (IIR) filters have an infinite impulse response, as they rely on feedback within the filter structure. IIR filters can

achieve higher filter order and sharper frequency selectivity but may introduce phase distortion.

2. Frequency Response:

The frequency response of a filter is a measure of how it responds to different frequencies of the input signal. It describes the filter's behavior as a function of frequency and is typically represented using a frequency response plot.

The frequency response of a filter is characterized by the following parameters:

- **Gain:** The gain of a filter is the ratio of the output signal amplitude to the input signal amplitude at a given frequency. It indicates the amplification or attenuation applied to the input signal.
- **Cutoff Frequency:** The cutoff frequency, also known as the corner frequency, is the frequency at which the filter starts to attenuate or pass signals. It determines the boundary between the passband and the stopband of a filter.
- **Passband:** The passband is the range of frequencies that the filter allows to pass with minimal attenuation. It is the desired frequency range that should be preserved.
- **Stopband:** The stopband is the range of frequencies that the filter attenuates or blocks significantly. It is the range of frequencies that should be suppressed or removed from the signal.
- **Filter Slope/Roll-off:** The filter slope or roll-off refers to how quickly the filter attenuates signals outside the passband or stopband. It indicates the rate at which the filter transitions from the passband to the stopband.

- **Phase Response:** The phase response of a filter describes the phase shift introduced by the filter as a function of frequency. It is important in applications where phase coherence is critical, such as in audio or communication systems.

Applications of Filters:

Filters have numerous applications across various fields, including:

- **Audio Systems:** Filters are used in audio systems for equalization, frequency shaping, noise reduction, and speaker protection.
- **Communication Systems:** Filters play a vital role in communication systems for signal modulation, demodulation, channel selection, and interference rejection.
- **Image Processing:** Filters are used in image processing applications for image enhancement, noise reduction, and edge detection.
- **Instrumentation:** Filters are employed in measurement and instrumentation systems for signal conditioning, noise rejection, and frequency analysis.
- **Biomedical Applications:** Filters are used in biomedical signal processing for electrocardiography (ECG), electroencephalography (EEG), and other physiological signal analysis.
- **Wireless Technologies:** Filters are utilized in wireless technologies such as Wi-Fi, cellular communication, and radar systems for frequency selection, interference rejection, and spectral shaping.

Conclusion:

Filters are essential components in electronic systems that shape the frequency characteristics of signals. They allow for the selection of desired frequencies and rejection of unwanted frequencies. Filters are available in various types, including passive filters, active filters, analog filters, and digital filters, each with its own advantages and applications. Understanding the concept of frequency response and the characteristics of filters is crucial for engineers working in fields such as audio systems, communication systems, image processing, and instrumentation.

CHAPTER-9: ELECTRONIC CIRCUITS AND DEVICES

Digital Logic Circuits

Digital logic circuits are the building blocks of digital systems and are composed of electronic components that perform logical operations on binary inputs to produce binary outputs. These circuits are the foundation of modern digital electronics and are used in a wide range of applications, including computers, calculators, digital signal processing, and control systems. This elaboration provides a detailed explanation of digital logic circuits, including their basic components, logic gates, and commonly used circuit configurations.

1. Basic Components of Digital Logic Circuits:

Digital logic circuits consist of several fundamental components that perform specific functions. The key components include:

- **Logic Gates:** Logic gates are elementary building blocks that perform basic logical operations, such as AND, OR, NOT, NAND, NOR, XOR, and XNOR. These gates take one or more binary inputs and produce a binary output based on the predefined logic function.
- **Flip-Flops:** Flip-flops are memory elements used to store binary data. They can be either edge-triggered or level-triggered and are essential for sequential logic circuits.

- **Registers:** Registers are composed of multiple flip-flops and are used to store and manipulate binary data. They are commonly used in applications requiring temporary storage, such as in microprocessors and data storage devices.
- **Decoders:** Decoders are used to convert a binary input code into a specific output pattern. They are often used in applications involving address decoding, memory systems, and display drivers.
- **Encoders:** Encoders perform the reverse function of decoders by converting multiple input signals into a coded binary output. They are used in applications such as data compression and multiplexing.
- **Multiplexers:** Multiplexers, also known as data selectors, are used to select one of many inputs and pass it to the output based on control signals. They are often used for data routing and selection purposes.
- **Demultiplexers:** Demultiplexers perform the opposite function of multiplexers by routing a single input to one of several outputs based on control signals.

2. Logic Gates:

Logic gates are the fundamental building blocks of digital logic circuits and perform specific logical operations on binary inputs to produce a binary output. There are several types of logic gates, including:

- **AND Gate:** The AND gate produces a high output (logic 1) only if all of its inputs are high (logic 1).
- **OR Gate:** The OR gate produces a high output (logic 1) if any of its inputs are high (logic 1).

- **NOT Gate:** The NOT gate, also known as an inverter, produces the complement of its input. If the input is high (logic 1), the output will be low (logic 0), and vice versa.
- **NAND Gate:** The NAND gate is an AND gate followed by a NOT gate. It produces a low output (logic 0) if all of its inputs are high (logic 1), and a high output (logic 1) otherwise.
- **NOR Gate:** The NOR gate is an OR gate followed by a NOT gate. It produces a low output (logic 0) if any of its inputs are high (logic 1), and a high output (logic 1) otherwise.
- **XOR Gate:** The XOR gate, or exclusive OR gate, produces a high output (logic 1) if the number of high inputs is odd; otherwise, it produces a low output (logic 0).
- **XNOR Gate:** The XNOR gate, or exclusive NOR gate, produces a high output (logic 1) if the number of high inputs is even; otherwise, it produces a low output (logic 0).

Logic gates can be combined in various ways to create more complex logic functions and circuits.

3. Common Digital Logic Circuit Configurations:

- **Combinational Logic Circuits:** Combinational logic circuits are composed of logic gates and other components interconnected to perform specific logical functions. These circuits produce an output based solely on the current input values and do not have any memory elements. Examples of combinational logic circuits include adders, multiplexers, encoders, and decoders.
- **Sequential Logic Circuits:** Sequential logic circuits utilize memory elements such as flip-flops

to store and manipulate binary data. These circuits have outputs that depend on both the current inputs and the past history of inputs. Sequential logic circuits can be synchronous or asynchronous. Synchronous circuits use a clock signal to synchronize the operations, while asynchronous circuits operate without a clock signal. Examples of sequential logic circuits include registers, counters, and finite state machines.

Circuit Diagrams:

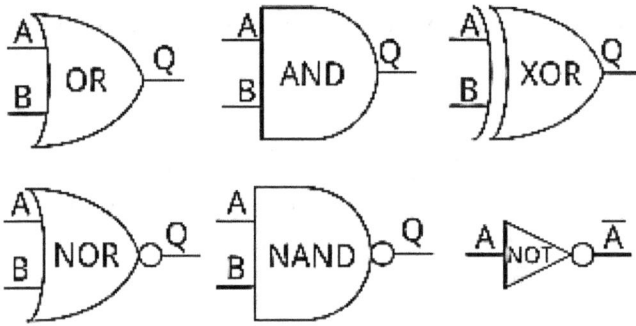

1. AND Gate:

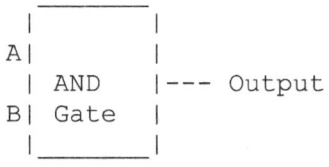

2. OR Gate:

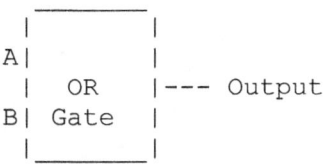

3. NOT Gate (Inverter):

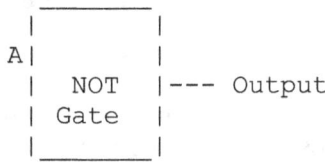

4. NAND Gate:

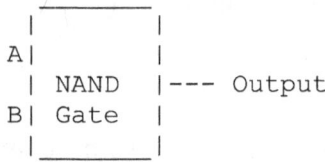

5. NOR Gate:

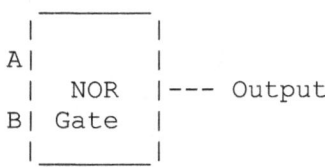

6. XOR Gate:

```
 _____
|        |
A|        |
 |  XOR   |--- Output
B|  Gate  |
 |_____|
```

These circuit diagrams illustrate the basic configurations of each logic gate. Inputs A and B are binary inputs, and the Output represents the resulting output based on the logic function of the gate.

Please note that these diagrams represent the logic gate symbols and their connections, but they do not include power supply connections or specific component values. In practical implementations, logic gates are often built using transistors or integrated circuits (ICs) that encapsulate the functionality of multiple gates in a single package.

Truth Tables:

Here are simplified diagrams representing some common digital logic circuits:

- **AND Gate:**

A	B	Output
0	0	0
0	1	0
1	0	0
1	1	1

- **OR Gate:**

A	B	Output
0	0	0
0	1	1
1	0	1
1	1	1

- **NOT Gate (Inverter):**

Input	Output
0	1
1	0

- **NAND Gate:**

A	B	Output
0	0	1
0	1	1
1	0	1
1	1	0

- **NOR Gate:**

A	B	Output
0	0	1
0	1	0
1	0	0
1	1	0

- **XOR Gate:**

```
A  B  Output
-------------------
0  0   0
0  1   1
1  0   1
1  1   0
```

These diagrams illustrate the logic behavior of the respective gates, showing the relationship between input values and the resulting output.

Conclusion:

Digital logic circuits form the basis of modern digital electronics, enabling the manipulation and processing of binary data. They are constructed using various components such as logic gates, flip-flops, registers, decoders, multiplexers, and demultiplexers. Logic gates perform logical operations on binary inputs, while combinational and sequential logic circuits combine these gates to create more complex functions. Understanding digital logic circuits and their components is essential for designing and implementing digital systems and computer architectures.

Analog and digital communication systems

Analog and digital communication systems are two distinct methods used to transmit information from one point to another. These systems play a crucial role in various fields, including telecommunications, broadcasting, data transmission, and networking. This elaboration provides a detailed explanation of analog and digital communication systems, highlighting their differences, advantages, and applications.

Analog Communication Systems:

Analog communication systems transmit information using continuous analog signals. In these systems, the signal varies in amplitude, frequency, or phase to represent the information being conveyed. The main components of analog communication systems include a transmitter, a channel, and a receiver.

1. Modulation: In analog communication systems, modulation is employed to transfer the information signal onto a carrier signal. The carrier signal has a much higher frequency than the information signal and is used to propagate the modulated signal through the channel.

- **Amplitude Modulation (AM):** Amplitude modulation is a technique where the amplitude of the carrier signal is varied in proportion to the instantaneous amplitude of the information signal.
- **Frequency Modulation (FM):** Frequency modulation is a technique where the frequency of the carrier signal is varied based on the instantaneous frequency of the information signal.
- **Phase Modulation (PM):** Phase modulation is a technique where the phase of the carrier signal is

varied according to the instantaneous phase of the information signal.

2. Advantages and Applications of Analog Communication Systems:

- **Wideband Transmission:** Analog communication systems are capable of transmitting a wide frequency range of signals, making them suitable for applications that require high-quality audio and video transmission, such as broadcast radio and television.
- **Simple Implementation:** Analog systems are often simpler to implement and require less complex signal processing compared to digital systems.
- **Natural Representation of Signals:** Analog systems provide a continuous representation of signals, making them well-suited for applications where the exact waveform or signal fidelity is crucial, such as in music and voice transmission.
- **Real-Time Communication:** Analog systems offer real-time communication capabilities, as the signal is continuously varying and can be instantaneously received and interpreted.

Digital Communication Systems:

Digital communication systems encode and transmit information using discrete binary signals, where the signal is represented by a sequence of 0s and 1s. These systems employ various modulation techniques and digital signal processing algorithms to ensure accurate transmission and reception of data.

1. Digital Modulation: In digital communication systems, modulation techniques are used to represent digital data in the form of binary signals.

- **Amplitude Shift Keying (ASK):** Amplitude Shift Keying modulates the amplitude of the carrier signal to represent digital data.
- **Frequency Shift Keying (FSK):** Frequency Shift Keying modulates the frequency of the carrier signal to represent digital data.
- **Phase Shift Keying (PSK):** Phase Shift Keying modulates the phase of the carrier signal to represent digital data.
- **Quadrature Amplitude Modulation (QAM):** QAM combines amplitude and phase modulation to transmit digital data using a constellation diagram.

2. Advantages and Applications of Digital Communication Systems:

- **Noise Immunity:** Digital communication systems are more resistant to noise and interference compared to analog systems. Digital signals can be accurately reconstructed even in the presence of noise.
- **Error Detection and Correction:** Digital communication systems allow for error detection and correction techniques, ensuring high data integrity.
- **Data Compression:** Digital systems facilitate efficient data compression techniques, enabling higher data transmission rates and storage capacity.
- **Multiple Access and Multiplexing:** Digital systems support multiple access techniques, allowing multiple users to share the same communication channel simultaneously.

- **Compatibility with Digital Devices:** Digital communication systems are compatible with digital devices, such as computers, smartphones, and digital networks, making them suitable for modern communication and networking technologies.

Conclusion

Analog and digital communication systems are distinct methods used to transmit information. Analog systems employ continuous analog signals and modulation techniques to transmit information, while digital systems utilize discrete binary signals and digital modulation techniques. Analog systems are suitable for applications requiring wideband transmission and natural representation of signals, while digital systems offer advantages such as noise immunity, error detection, and compatibility with digital devices. Both analog and digital communication systems are widely used in various fields, and their selection depends on specific requirements, bandwidth limitations, and the nature of the information being transmitted.

CHAPTER-10: INTRODUCTION TO ELECTRICAL MACHINES

DC machines

DC machines, also known as direct current machines, are electromechanical devices that convert electrical energy into mechanical energy or vice versa. They are widely used in various applications such as electric vehicles, industrial machinery, generators, and motors. This elaboration provides a detailed explanation of DC machines, including their construction, working principles, types, and applications.

1. Construction of DC Machines:

DC machines consist of two main parts: the stator (field system) and the rotor (armature). The stator is the stationary part of the machine and is responsible for creating a magnetic field, while the rotor is the rotating part that carries the armature windings.

The stator is typically made up of a set of permanent magnets or electromagnetic coils, called field coils, which produce the magnetic field. The rotor, on the other hand, consists of a cylindrical core made of laminated iron, with armature windings wound around it. The armature windings are connected to a commutator, which is a segmented cylindrical conductor, and the commutator is in contact with brushes that allow the electrical connection with the external circuit.

2. Working Principles of DC Machines:

DC machines operate based on the principles of electromagnetic induction and the interaction between the magnetic field and the armature windings.

- **Generator Mode:** In the generator mode, when mechanical energy is supplied to the rotor by an external source, the armature windings cut through the magnetic field, inducing an electromotive force (EMF) according to Faraday's law of electromagnetic induction. This generated EMF can be tapped from the brushes and used as electrical energy.
- **Motor Mode:** In the motor mode, when electrical energy is supplied to the armature windings through the brushes, a current flows through the windings, creating a magnetic field. This magnetic field interacts with the magnetic field produced by the stator, resulting in a mechanical force that rotates the rotor.

3. Types of DC Machines:

There are two main types of DC machines: DC generators and DC motors.

- **DC Generators:** DC generators convert mechanical energy into electrical energy. They operate in the generator mode, where the rotational motion of the rotor induces an EMF in the armature windings, generating a DC voltage output.
- **DC Motors:** DC motors convert electrical energy into mechanical energy. They operate in the motor mode, where the current flowing through the armature windings interacts with the magnetic field, producing a torque that rotates the rotor.

DC generators and motors can further be classified based on their excitation methods and connections, such as shunt-wound, series-wound, compound-wound, separately excited, or permanent magnet types.

4. Applications of DC Machines:

DC machines find applications in various industries and fields due to their reliable performance and controllable characteristics. Some common applications include:

- Electric Vehicles: DC motors are used in electric vehicles for propulsion, providing the necessary torque and speed control.
- Industrial Machinery: DC machines are employed in industrial machinery, such as conveyor systems, pumps, fans, and machine tools, where precise speed control is required.
- Generators: DC generators are used in power plants, renewable energy systems, and portable generators to produce DC power.
- Robotics: DC motors are widely used in robotic systems for precise motion control and positioning.
- Battery Charging: DC generators are used to charge batteries in applications like backup power systems, mobile charging stations, and battery-operated equipment.

Conclusion:

DC machines are versatile devices that convert electrical energy into mechanical energy or vice versa. They are constructed with a stator (field system) and a rotor (armature) and operate based on electromagnetic induction. DC generators convert mechanical energy into electrical energy, while DC motors convert electrical energy into mechanical energy. They find applications in electric vehicles, industrial machinery, generators, robotics, and battery charging systems. DC machines continue to play a significant role in various industries, providing efficient and controllable power conversion solutions.

Induction motors

Induction motors, also known as asynchronous motors, are widely used in various industrial and commercial applications for converting electrical energy into mechanical energy. They are robust, reliable, and efficient, making them the most commonly used type of electric motor. This elaboration provides a detailed explanation of induction motors, including their construction, working principles, types, and applications.

1. Construction of Induction Motors:

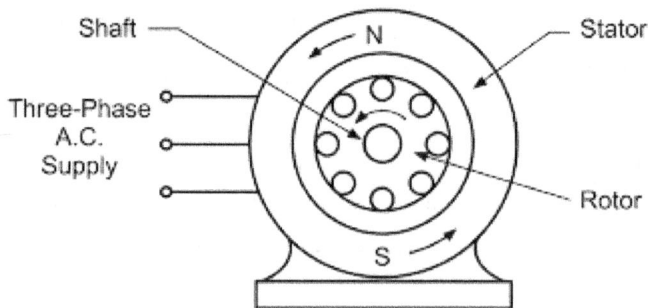

Induction motors consist of two main parts: the stator and the rotor.

- **Stator:** The stator is the stationary part of the motor and consists of a laminated core made of magnetic materials, such as silicon steel. It is designed with slots to accommodate stator windings. The stator

windings are typically connected in a three-phase configuration and are supplied with an alternating current (AC).
- **Rotor:** The rotor is the rotating part of the motor and is made of a laminated core with slots to hold rotor conductors. The rotor conductors are typically made of aluminum or copper bars, short-circuited at the ends by end rings. The rotor can be either squirrel cage or wound type.

2. Working Principles of Induction Motors:

Induction motors work based on the principles of electromagnetic induction and the interaction between the stator and rotor magnetic fields.

- **Stator Magnetic Field:** When three-phase AC current is supplied to the stator windings, a rotating magnetic field is produced. The rotating magnetic field rotates at a synchronous speed, which depends on the frequency of the AC supply and the number of poles in the stator.
- **Rotor Interaction:** The rotating magnetic field induces a current in the rotor conductors through electromagnetic induction. This induced current creates a magnetic field in the rotor. The interaction between the stator magnetic field and the rotor magnetic field produces a torque that causes the rotor to rotate.
- **Slip:** The actual speed of the rotor is slightly less than the synchronous speed, resulting in slip. The slip is necessary for the motor to generate torque and provide mechanical work.

3. Types of Induction Motors:

There are two main types of induction motors: squirrel cage and wound rotor motors.

- **Squirrel Cage Induction Motors:** Squirrel cage motors have a rotor with short-circuited rotor bars and end rings, resembling a squirrel cage. They are simple in construction, reliable, and widely used in applications where high starting torque is not required.

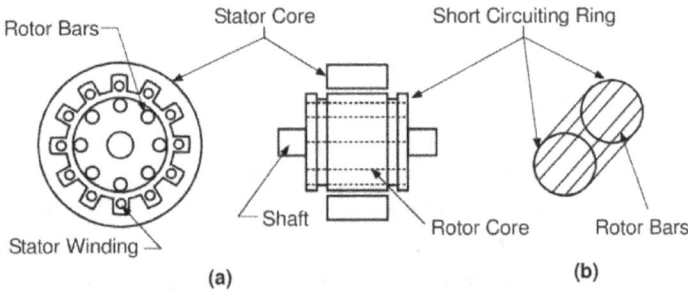

- **Wound Rotor Induction Motors:** Wound rotor motors have a rotor with windings connected to external resistors or slip rings. This allows for additional control of rotor impedance and provides the ability to vary the speed and torque characteristics of the motor. Wound rotor motors are suitable for applications requiring high starting torque, adjustable speed, and regenerative braking.

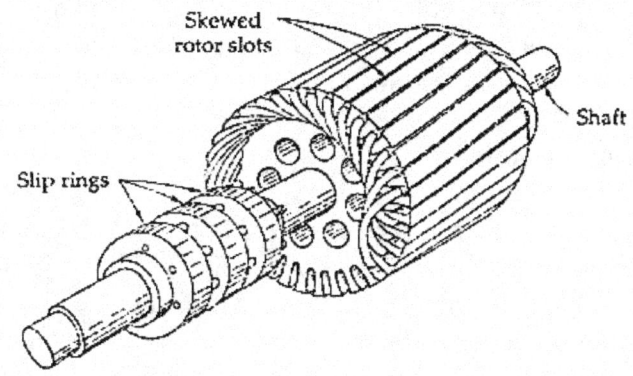

4. Applications of Induction Motors:

Induction motors are extensively used in various industries and applications due to their robustness, simplicity, and efficiency. Some common applications include:

- Industrial Machinery: Induction motors are used in pumps, compressors, fans, conveyors, mixers, and other industrial equipment that require reliable and efficient motor operation.
- HVAC Systems: Induction motors power the fans and compressors in heating, ventilation, and air conditioning (HVAC) systems, providing cooling and air circulation.
- Home Appliances: Induction motors are found in appliances like refrigerators, washing machines, air conditioners, and kitchen appliances, enabling their efficient operation.
- Automotive Industry: Induction motors are used in electric and hybrid vehicles for propulsion, providing high torque and efficiency.

- Renewable Energy Systems: Induction generators are used in wind turbines and hydroelectric systems to convert mechanical energy into electrical energy.

Conclusion:

Induction motors are the workhorses of the industrial and commercial sectors, providing efficient and reliable mechanical power. They are constructed with a stator and rotor and operate based on the principles of electromagnetic induction. Induction motors are available in squirrel cage and wound rotor types, each suited for specific applications. They find widespread use in industrial machinery, HVAC systems, home appliances, automotive industry, and renewable energy systems. The versatility, efficiency, and durability of induction motors make them a fundamental component of modern society.

Synchronous Machines

Synchronous machines are a type of alternating current (AC) electrical machine that operate at a fixed speed, synchronized with the frequency of the electrical grid. They are widely used in power generation, industrial applications, and high-performance motor drives. This elaboration provides a detailed explanation of synchronous machines, including their construction, working principles, types, and applications.

1. Construction of Synchronous Machines:

Synchronous machines consist of two main parts: the stator and the rotor.

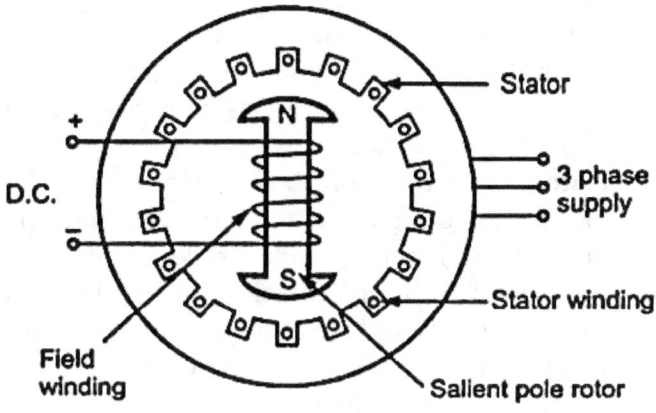

- **Stator:** The stator is the stationary part of the machine and contains the stator windings. The stator windings are usually three-phase windings placed in slots on the stator core. These windings produce a rotating magnetic field when energized by an AC power supply.
- **Rotor:** The rotor is the rotating part of the machine and can be constructed in different configurations depending on the type of synchronous machine. The rotor contains field windings or permanent magnets that produce a magnetic field. The rotor is designed to maintain synchronization with the rotating magnetic field produced by the stator.

2. Working Principles of Synchronous Machines:

Synchronous machines operate based on the principle of electromagnetic induction and maintain synchronism with the AC power supply frequency.

- **Stator Magnetic Field:** When three-phase AC current is supplied to the stator windings, a rotating

magnetic field is generated. The rotating magnetic field rotates at a constant speed, known as the synchronous speed, which is determined by the frequency of the AC power supply and the number of poles in the stator.
- **Rotor Synchronization:** In synchronous machines, the rotor is designed to rotate at the same speed as the rotating magnetic field produced by the stator. This synchronization is achieved by either using a DC current to produce a magnetic field in the rotor windings or using permanent magnets.
- **Torque Generation:** When the rotor is synchronized with the rotating magnetic field, a torque is developed in the machine. This torque can be used to drive mechanical loads or to generate electrical power in synchronous generators.

3. Types of Synchronous Machines:

There are two main types of synchronous machines: synchronous generators (alternators) and synchronous motors.

- **Synchronous Generators:** Synchronous generators convert mechanical energy into electrical energy. They are commonly used in power plants to generate electricity. The rotor of a synchronous generator is typically excited with a DC current to create a magnetic field.

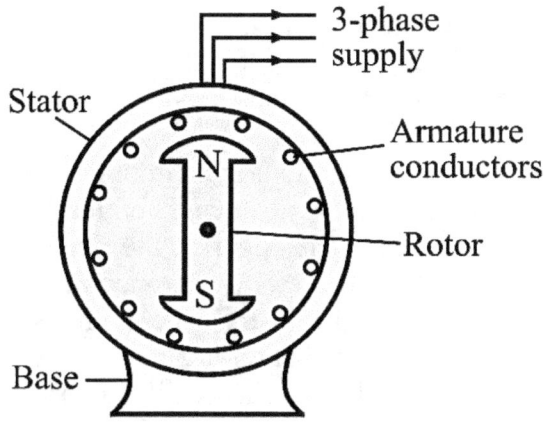

- **Synchronous Motors:** Synchronous motors convert electrical energy into mechanical energy. They are used in applications that require constant speed and high precision, such as industrial drives, pumps, compressors, and synchronous clocks. Synchronous motors can have rotor windings or permanent magnets to produce the magnetic field.

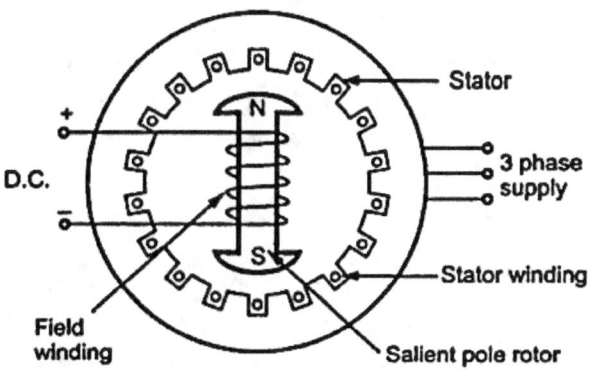

-

4. **Applications of Synchronous Machines:**

Synchronous machines have numerous applications in different industries due to their unique characteristics. Some common applications include:

- Power Generation: Synchronous generators are used in power plants to generate electricity and provide reactive power support to the electrical grid.
- Industrial Drives: Synchronous motors are used in industrial machinery where precise speed control and synchronization are required, such as in paper mills, steel mills, and textile factories.
- Renewable Energy Systems: Synchronous generators are used in wind turbines and hydroelectric power plants to convert mechanical energy into electrical energy from renewable sources.
- High-Performance Motor Drives: Synchronous motors are used in high-performance motor drives, such as robotics, CNC machines, and electric vehicle propulsion systems, where precise control and high efficiency are crucial.
- Synchronous Clocks: Synchronous motors are used in synchronous clocks and timing devices, ensuring accurate timekeeping.

Conclusion:

Synchronous machines play a vital role in various industries and applications. They are constructed with a stator and rotor and operate based on the principles of electromagnetic induction and synchronization with the AC power supply frequency. Synchronous generators convert mechanical energy into electrical energy, while synchronous motors convert electrical energy into mechanical energy. They find applications in power generation, industrial drives, renewable energy systems, high-performance motor drives, and timing devices. The precise control, constant speed, and synchronization capabilities of synchronous machines make them essential components in modern electrical systems.

Transformers in Power Systems

Transformers are vital components in electrical power systems used for the efficient transmission, distribution, and utilization of electrical energy. They provide a means of changing voltage levels while maintaining power frequency, allowing for efficient power transfer and voltage regulation. This elaboration provides a detailed explanation of transformers, including their construction, working principles, types, and applications in power systems.

1. Construction of Transformers:

Transformers consist of two or more coils of wire wound around a laminated iron core. The two main coils are known as the primary winding and the secondary winding. The primary winding is connected to the power source, while the secondary winding is connected to the load. The laminated core is made of iron to provide a low-reluctance path for the magnetic flux.

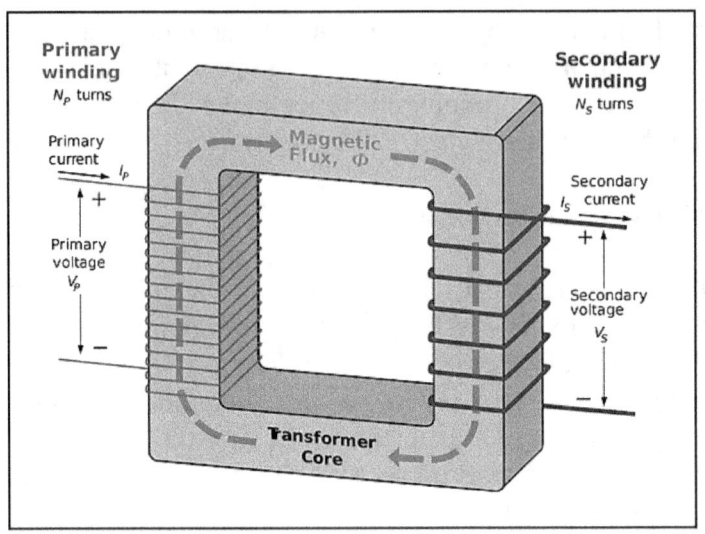

2. Working Principles of Transformers:

Transformers operate based on the principle of electromagnetic induction. When an alternating current (AC) flows through the primary winding, it creates a changing magnetic field in the core. This changing magnetic field induces a voltage in the secondary winding through electromagnetic induction. The ratio of the number of turns in the primary winding to the number of turns in the secondary winding determines the voltage transformation ratio.

3. Types of Transformers:

There are several types of transformers used in power systems, including:

- **Step-Up Transformers:** Step-up transformers increase the voltage level from the primary winding to the secondary winding. They are used in power

generation stations to step up the voltage for long-distance transmission to reduce power losses.
- **Step-Down Transformers:** Step-down transformers decrease the voltage level from the primary winding to the secondary winding. They are used in distribution systems to lower the voltage for safer utilization in residential, commercial, and industrial areas.
- **Auto Transformers:** Auto transformers have a common winding shared by the primary and secondary circuits. They are used for voltage adjustment and can provide a range of voltage transformations.
- **Isolation Transformers:** Isolation transformers provide electrical isolation between the primary and secondary windings. They are commonly used for safety purposes to isolate sensitive equipment from the power supply and to reduce noise and interference.
- **Instrument Transformers:** Instrument transformers, such as current transformers and potential transformers, are used for measuring and protection purposes. They scale current and voltage levels to a standard value suitable for measuring instruments and protective relays.

4. Applications of Transformers:

Transformers are used extensively in power systems for various applications, including:

- **Transmission and Distribution:** Transformers are used in high-voltage transmission networks to step up the voltage for efficient long-distance power transmission and step it down at substations for distribution to lower voltage levels.

- **Power Generation:** Transformers are used in power plants to step up the generated voltage for transmission to the grid.
- **Industrial Applications:** Transformers are used in industrial settings for supplying power to machinery, motors, and equipment at appropriate voltage levels.
- **Residential and Commercial Buildings:** Transformers are used in buildings for voltage transformation and distribution of electrical power to meet the requirements of lighting, appliances, and other electrical loads.
- **Renewable Energy Systems:** Transformers are used in renewable energy systems, such as solar and wind farms, to step up the generated voltage for connection to the power grid.
- **Railway Systems:** Transformers are used in railway systems to supply power to electric locomotives and trains.

Conclusion:

Transformers are essential components in power systems, enabling efficient transmission, distribution, and utilization of electrical energy. They operate based on the principle of electromagnetic induction and provide voltage transformation while maintaining power frequency. With their various types and applications, transformers play a crucial role in ensuring reliable and safe power supply for residential, commercial, industrial, and infrastructure sectors.

Basic Motor Control

Motor control refers to the process of starting, stopping, and regulating the speed and direction of electric motors. It involves the use of various control devices and techniques to manage the operation of motors in industrial, commercial, and residential applications. This elaboration provides a detailed explanation of basic motor control, including the components, control methods, and applications.

1. Components of Motor Control Systems:

Motor control systems consist of several key components that work together to control the operation of electric motors:

- **Power Supply:** The power supply provides the electrical energy necessary to operate the motor. It can be a direct current (DC) or alternating current (AC) power source, depending on the motor type.
- **Motor Controller:** The motor controller is the central component that manages the operation of the motor. It receives input signals and provides output signals to control the motor's speed, direction, and other parameters.
- **Control Devices:** Control devices include switches, push buttons, relays, contactors, and motor starters. They are used to manually or automatically control the motor's operation, such as starting, stopping, and reversing.
- **Sensors:** Sensors, such as temperature sensors, current sensors, and speed sensors, are used to monitor the motor's performance and provide feedback to the motor controller. This feedback is essential for implementing control strategies and ensuring safe and efficient motor operation.

2. Control Methods in Motor Control:

There are several control methods used in motor control systems, depending on the application requirements and motor type:

- **Direct On/Off Control:** In this control method, the motor is directly switched on and off using a switch or contactor. It provides basic control but lacks speed regulation and precision.
- **Reversing Control:** Reversing control allows for changing the direction of motor rotation. It involves the use of reversing starters or reversing contactors to switch the motor's connections and reverse the current flow.
- **Speed Control:** Speed control methods allow for regulating the motor's speed. They can include voltage control, where the motor's voltage is varied, or frequency control, where the motor's frequency is adjusted.
- **Soft Start and Stop Control:** Soft start and stop control gradually ramps up the motor's voltage or frequency during startup and shutdown to reduce mechanical stress and power surges.
- **Closed-Loop Control:** Closed-loop control utilizes feedback from sensors to monitor the motor's performance and adjust the control signals accordingly. It allows for precise control of motor speed, torque, and position.

3. Applications of Motor Control:

Motor control is applied in various industries and applications to manage the operation of electric motors effectively. Some common applications include:

- **Industrial Machinery:** Motor control is extensively used in industrial machinery, such as pumps, compressors, conveyor systems, and manufacturing equipment, to ensure precise operation and optimize energy efficiency.
- **HVAC Systems:** Motor control is employed in heating, ventilation, and air conditioning (HVAC) systems to control fans, blowers, and pumps for efficient air circulation and temperature regulation.
- **Home Appliances:** Motor control is utilized in household appliances, including refrigerators, washing machines, dishwashers, and HVAC units, to control their motor-driven components and optimize their performance.
- **Automotive Industry:** Motor control plays a critical role in the automotive industry for controlling electric motors in electric vehicles, hybrid vehicles, and various vehicle subsystems, such as power windows, wipers, and seat adjustments.
- **Robotics and Automation:** Motor control is essential in robotics and automation systems to control the precise movement and positioning of robotic arms, actuators, and other motor-driven components.
- **Renewable Energy Systems:** Motor control is employed in renewable energy systems, such as wind turbines and solar tracking systems, to optimize energy capture, regulate power output, and ensure efficient operation.

Conclusion:

Basic motor control systems are fundamental in managing the operation of electric motors in a wide range of applications. They involve the use of components such as motor controllers, control devices, and sensors to control the motor's speed, direction, and other parameters. By employing various control methods, motor control systems ensure efficient, safe, and precise operation of electric motors in industrial, commercial, and residential settings.